Lecture Notes in Chemistry

Edited by G. Berthier, M. J. S. Dewar, H. Fischer
K. Fukui, H. Hartmann, H. H. Jaffé, J. Jortner
W. Kutzelnigg, K. Ruedenberg, E. Scrocco, W. Zeil

12

The Permutation Group in Physics and Chemistry

Edited by Jürgen Hinze

Springer-Verlag
Berlin Heidelberg New York 1979

Editor

Jürgen Hinze
Fakultät für Chemie
Universität Bielefeld
Universitätsstr.
Postfach 8640
4800 Bielefeld

ISBN 3-540-09707-4 Springer-Verlag Berlin Heidelberg New York
ISBN 0-387-09707-4 Springer-Verlag New York Heidelberg Berlin

Library of Congress Cataloging in Publication Data
Hinze, Jürgen, 1937-
The permutation group in physics and chemistry
(Lecture notes in chemistry; v. 12)
Bibliography: p.
Includes index.
1. Chemistry, Physical and theoretical--Mathematics.
2. Permutation groups. I. Title.
QD455.3.M3H56 541'.2'015122 79-23305
ISBN 0-387-09707-4

Printing and binding: Beltz Offsetdruck, Hemsbach/Bergstr.
2152/3140-543210

Table of Contents

Introduction

The permutation group has gained prominence in the fundamental research in diverse areas of physics and chemistry. Covering all salient developments of the last few years in a single symposium would require weeks, legions of participants and parallel sessions, highlighting the differences in language and communication problems between pure mathematicians, high and low energy physicists and chemists. The symposium held July 1978 at the Centre of Interdisciplinary Studies of the University of Bielefeld focussed on a small area, the pertinence of the permutation group in chemical physics, with the goal to increase and generate a fruitful dialogue between mathematicians and chemists.

In chemistry, concerned with the electronic and geometric structure of molecules as well as elementary chemical reactions, i.e. rearrangements in these structures, the permutation group has its relevance, since with its representations the effects and consequences of exchanging indistinguishable particles, electrons and identical nuclei, can be systematized and classified. This may be exemplified by a brief survey of the lectures presented, which may also serve as a first orientation to the articles of this volume. In the first two contributions by A. Kerber and J.G. Nourse, the permutation group is used in the counting and systemtaic generation of stereoisomers aiding in the elucidation of possible molecular structures. The dynamics of stereochemistry is considered in the next article by J.G. Nourse. The following section with four contributions by P.R. Bunker, J.D. Louck, A. Dress and R.S. Berry deals with the permutation group as a generalization of the point group symmetry useful in the interpretation of molecular spectra and structure. This generalization originally suggested by Longuet-Higgings, who unfortunately could not be present at the symposium, provides a frame for the concise description of molecular dynamics with minimal coupling between internal and external motion and is of particular importance when nonrigid molecules are considered. How far this concept can lead is illustrated in the contribution by R.S.Berry, where the full transition of structured molecules to liquid drops is investigated; possibly a guide to a symmetry classification of molecular rearrangements and the kinematics of chemical reactions. The next section with three articles by L.C. Biedenharn, J.D. Louck and B.R. Judd, deal with the representation and classification of electronic wavefunctions for molecules and atoms using the irreducible representations of the symmetric group as a basis for the description of the spin and angular momentum coupling. How such a symmetry basis can be used effectively for the evaluation of matrix elements, important in the quantitative description of electronic and molecular structure, is illustrated in the contributions by T.H. Seligman and J.S. Frame. This is related to the more recently developed methods for the rapid calculation of configuration mixing matrix elements using the unitary group approach, a topic for a forthcoming special symposium. A final contribution by A. Dress comments on some mathematical aspects of the chirality algebra of Ruch, who was present at and contributed to the symposium.

Finally I would like to use this opportunity to acknowledge the assistance of the organizing committee in the planning of the symposium and express my gratitude to the directorship and staff of the Centre for Interdisciplinary Studies for the financial and organizational assistance, which made the symposium possible. My special thanks go to Mrs. K. Mehandru and B. Yurtsever for the typing of the manuscripts.

Bielefeld, August 1979

The editor:

Juergen Hinze

Organizing Committee: A. Dress
J. Hinze
E. Ruch

Counting isomers and Such

A. Kerber

Lehrstuhl D für Mathematik
RWTH Aachen
Templergrabel 64
5100 Aachen, Germany

This conference is devoted to applications of permutation groups in chemistry and physics. Hearing of applications of group theory in sciences, our first reaction is to think of symmetry considerations, where the situation is as follows: a given system is symmetric with respect to a certain symmetry group, a subgroup, say, of the three-dimensional orthogonal group, and this invariance of the given system under symmetry operations can be used in order to attack the corresponding mathematical problem.

I would like to describe in this introductory lecture a slightly different kind of problem where group theory can be applied: The counting of isomeric molecules and related topics. In this situation group theory is used in order to first define and then count a certain structure (namely isomorphism types of graphs) as equivalence classes on certain sets, which again is a typical and very important mathematical procedure.

Another reason for me to choose this topic was that this chemical problem of counting isomers gave rise to two important mathematical theories namely to both the graph theory and the theory of enumeration under group action (which is now an important part of combinatorics) (for more details see ref.(1)). The aim is to give a review over the present status of this theory. The main emphasis lies on the crucial lemmata of Burnside and deBruijn which yield the tools for calculations.

1. The history

By the middle of the nineteenth century chemists began to develop various kinds of notations in order diagrammatically to represent molecules. The method which is essentially the one in use today was introduced in 1864 by the Edinburgh chemist Alexander Crum Brown (see ref. (2)), who represented each atom separately by a letter enclosed in a circle and all single and multiple bonds by lines joining the circles. For example CO_2 was represented by

1.1  .

Since in this case there is exactly one of the three atoms in the molecule, which is of valency 4, while the other ones both have valency 2, we may very well simplify the notation in order to represent CO_2 by

1.2  ,

a structure for which J.J. Sylvester in his paper "Chemistry and Algebra" (ref. (3)) introduced the name graph. When Crum Brown introduced his notation there was a strange and unexplained fact called isomerism, substances had been found which had the same empirical formula but different properties. The success of his notation is at least partly due to the fact that it explained isomerism very well by showing that it is caused by the possibilities to arrange the same atoms in different ways, i.e. in different graphs. For example C_4H_{10} has two isomers, the graphs of which are

1.3 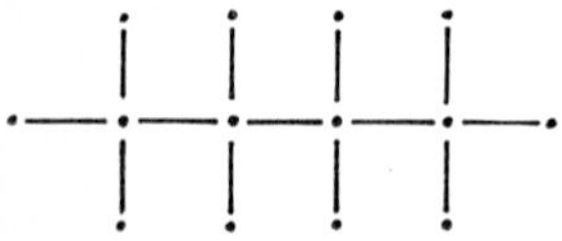 and 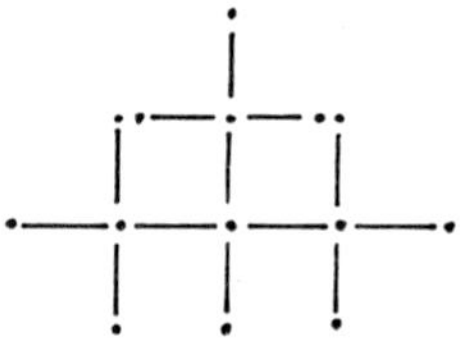 .

In 1874 A. Cayley wrote a paper (ref.(4)) which points to the connection of his work on the enumeration of trees and this problem of counting the number of chemical isomers, so that the problem of isomerism was now put into a mathematical framework.

Various papers were written, many particular results were derived, but it was only in 1929 when a joint paper of A.C. Lunn (a chemist) and J.K Senior (a mathematician) appeared (ref.(5)), which clearly recognizes that the theory of permutation groups is the appropriate tool in the theory of enumeration of isomers, so that a systematic treatment was started. The final breakthrough is due to G. Pólya and his excellent paper entitled "Kombinatorische Anzahlbestimmungen für Gruppen, Graphen und chemische Verbindungen" (ref. (6)). In the fifties of this century, graph theorists used these enumerative methods developed by Pólya and others and solved many enumeration problems in their field of interest. In this way enumeration theory made a rapid progress and today it is an independent and important part of combinatorics, the first book on this subject came out recently (ref.(7)). In a sense it amounts to a theory of Burnside's lemma on the number of orbits of permutation groups and of generalizations of this lemma. The methods are therefore algebraic ones. I would like to give a short review of the present situation of this theory. In order to do this, a precise formulation of its basic problems will be in order.

2. Basic problems

We remember that the enumeration theory arose from the problem of counting the isomers of a given empirical formula like C_4H_{10}. This means that we ask for the number of types of graphs with 14 points, 4 of which are of valency 4 (so that they correspond to the carbon atoms) while 10 of them are of valency 1. Hence more generally we ask in fact for the number of multigraphs with prescribed edge degree sequence as it will be described next. A multigraph consists of points $p_1, p_2, \ldots$ and edges $e_1, e_2, \ldots$ like for example

2.1

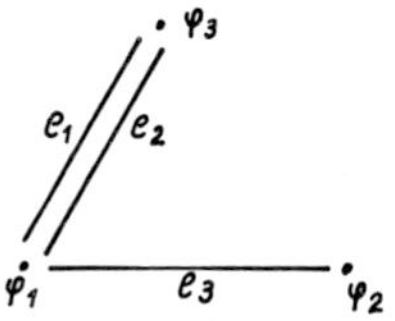

But this is not exactly what we want to consider since we shall neglect the labeling, so that from 2.1 we get the following structure:

2.2

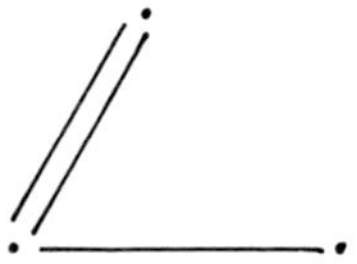

In order to make this more precise we have to express all this in mathematical terms.

First of all we define a multigraph with p points and up to k-fold edges to a mapping γ from the set $\{1,\ldots,\binom{p}{2}\}$ (which is interpreted as being the set of the $\binom{p}{2} = \frac{p(p-1)}{2}$ pairs of points) into the set $\{0,1,\ldots,k\}$ of multiplicities (of the edges, so that $\gamma(i) = j$ means that the i-th pair of points is connected by a j-fold edge), for short:

2.3 $$\gamma : \{1,\ldots,\binom{p}{2}\} \to \{0,\ldots,k\} .$$

Or (if we denote by Y^X the set of all the mappings from X into Y):

2.4 $$\gamma \in \{0,\ldots,k\}^{\{1,\ldots,\binom{p}{2}\}}.$$

Hence there exist $(k+1)^{\binom{p}{2}}$ multigraphs with p points and up to k-fold edges.

In order to obtain from this whole lot of multigraphs (like 2.1) the number of types of multigraphs (like 2.2) we introduce an equivalence relation on the set of multigraphs such than an element of an equivalence class gives just such a type by neglecting the labeling. We define this relation by saying that two such multigraphs γ and γ' are *equivalent* or *isomorphic* if and only if γ' arises from γ by just permuting the points. Permuting the points means that there exists a permutation π of the p points, i.e. $\pi \in S_p$ (the symmetric group which consists of all the permutations of the set $\{1,\ldots,p\}$), which satisfies

2.5 $$\gamma' = \gamma \circ \pi^{[2]-1}$$

(i.e. $\gamma'(i) = \gamma(\pi^{[2]-1}(i))$) for the permutation $\pi^{[2]}$ (and its inverse $\pi^{[2]-1}$) which is induced by π on the set $\{1,\ldots,\binom{p}{2}\}$ of the pairs of points.

In this way we have defined a permutation group

2.6 $$S_p^{[2]} = \{\pi^{[2]} \mid \pi \in S_p\},$$

which acts on the set 2.4 in such a way that each one of the equivalence classes consisting of isomorphic multigraphs is just an *orbit* of this group. A representative of the corresponding *type* of multigraphs is obtained by taking any element of the orbit in question and neglecting the labeling. For example if $k = 1$ and $p = 3$, we obtain the following types:

2.7 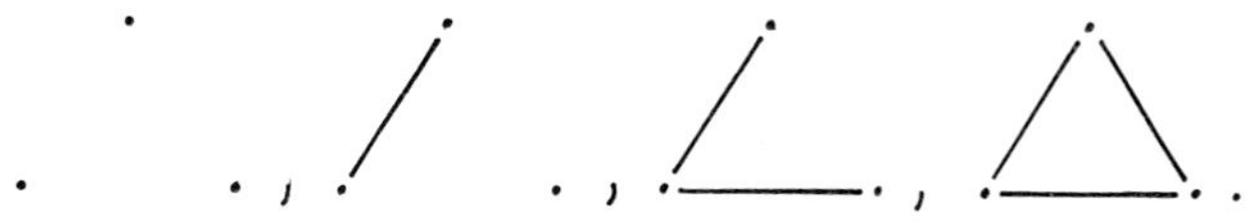

The *edge degree* of a point of a multigraph or of such an isomorphism type of multigraphs is the number of edges which meet in this point. Taking the edge degrees in their natural order, we obtain the edge degree sequence of the graph in question. The edge degree sequence of 2.1 and 2.2 for example is (1,2,3).

It is now clear from the foregoing that counting isomers amounts to evaluating the numbers of the types of multigraphs with prescribed edge degree sequence.

This problem which was the starting point for enumeration theory can be generalized very easily as follows. Instead of the set $\{0,\ldots,k\}$ of the multiplicities of edges we consider any two finite sets, say X and Y and together with these the set

2.8 $$Y^X = \{\varphi \mid \varphi : X \to Y\}$$

consisting of all the mappings from X into Y. Instead of the group $S_p^{[2]}$ acting on $\{1,\ldots,\binom{p}{2}\}$ we consider an arbitrary permutation group H acting on X, and besides this we assume that also a permutation group G acting on Y is given.

G and H give rise to the following equivalence relations "$\sim_i$", i = 1,2,3, on the set Y^X:

2.9 (i) $\varphi \sim_1 \psi \iff \exists h \in H \quad \forall x \in X(\varphi(x) = \psi(h^{-1}(x)))$,

(ii) $\varphi \sim_2 \psi \iff \exists h \in H, \quad g \in G \forall x \in X(\varphi(x)=g(\psi(h^{-1}(x))))$,

(iii) $\varphi \sim_3 \psi \iff \exists h \in H, \quad f:X \to G\ x \in \forall X(\varphi(x)=f(x)\psi(h^{-1}(x)))$.

The basic problems of enumeration theory now read as follows:

2.10 (i) How many equivalence classes has "$\sim_i$"?

(ii) How many of the equivalence classes consist of elements φ with certain properties (like given content $|\varphi^{-1}[\{y\}]|$, $y \in Y$)?

(iii) How can we obtain a complete system of representatives of the classes?

Refering to graphical enumeration we can put for an example $X = \{1,\ldots,\binom{p}{2}\}$, $Y = \{0,1,\ldots,k\}$ and $H = S_p^{[2]}$ in which case an answer to 2.10 (i) - (ii) yields

2.11 (i) The number of nonisomorphic multigraphs with p points and up to k-fold edges.

(ii) The number of nonisomorphic multigraphs with p points and up to k-fold edges, and prescribed number of i-fold edges, $0 \leq i \leq k$.

(iii) A complete system of types of multigraphs with p points and up to k-fold edges.

3. Solution methods

2.10 can be solved by an application of Burnside's lemma, which reads as follows:

3.1 Burnside's lemma:

If M is a finite set and $\delta: A \to S_M$ a permutation representation of a finite group A on M (i.e. $\delta(aa') = \delta(a)\delta(a')$ and each $\delta(a)$ an element of the symmetric group S_M on M) then the number of orbits of $\delta[A]$ on M is equal to

$$\frac{1}{|A|} \sum_{\alpha \in A} a_1(\delta(\alpha)) = \frac{1}{|\delta[A]|} \sum_{\sigma \in \delta[A]} a_1(\sigma)$$

if $a_1(\delta(\alpha))$ denotes the number of elements $m \in M$ which remain fixed under $\delta(\alpha)$ (i.e. $\delta(\alpha)(m) = m$).

Hence in order to solve 2.10 (i) it remains to recognize the classes of $\sim_i$ as orbits of a permutation group and to evaluate the number of fixed points for each element of this permutation group.

In the case when we are dealing with multigraphs again, we have for each $\pi^{[2]} \in S_p^{[2]}$:

3.2 $$a_1(\delta(\pi^{[2]})) = (k+1)^{c(\pi^{[2]})},$$

where $c(\pi^{[2]})$ denotes the number of cyclic factors of $\pi^{[2]}$. Hence the number of multigraphs on p points with up to k-fold edges is equal to

3.3 $$\frac{1}{p!} \sum_{\pi \in S_p} (k+1)^{c(\pi^{[2]})}.$$

Problem 2.10 (ii), the enumeration of the number of classes with prescribed properties, can be attacked by an application of a generalization of 3.1 which is called the "weighted form" of Burnside's lemma. In order to describe it, we introduce weights as follows. A mapping w from a set M into the ring $\mathbb{Q}[z_1, z_2, \ldots]$ of all the polynomials in the indeterminates $z_1, z_2, \ldots$ (we shall take suitably many of them) is called a <u>weight function</u>.

3.4 Weighted form of Burnside's lemma:

If **M**, A and δ are like in 3.1, and if $w:\mathbf{M} \to \mathbb{Q}[z_1,z_2,\ldots]$ is a weight function which is constant on the orbits ω of $\delta[A]$ on **M**, and if we define the weight w_ω of the orbit ω by $w(\varphi)$, φ ω, then we have for the sum of the weights of the orbits:

$$\sum_{\omega} w_\omega = \frac{1}{|A|} \sum_{\alpha \in A} \sum_{\alpha(m) = m} w(m).$$

In order to provide an example we consider "$\sim_1$" and take the following weight function. Any mapping $w^*:Y \to \mathbb{Q}[z_1,\ldots,_{|Y|}]$, defined by

3.5 $$w^*(y_i) = z_i \ ,$$

for a fixed ordering $Y = [y_1,\ldots,y_{|Y|}]$ of Y yields the weight function $w:Y^X \to \mathbb{Q}[z_1,\ldots,z_{|Y|}]$ by putting

3.6 $$w(\varphi) = \prod_x w(\varphi(x)).$$

Recalling that $\varphi \in Y^X$ remains fixed under $h \in H$ if and only if φ is constant on the elements of each cyclic factor of h and that

(if $h = \prod_{\nu=1}^{c(h)} (j_\nu h(j_\nu)\ldots h^{l_\nu - 1}(j_\nu))$ is the cycle decomposition of h)

such a φ has the following weight:

3.7 $$\prod_{\nu=1}^{c(h)} z^{l_\nu}_{(j_\nu)} \ ,$$

then we get by 3.4: If $a_k(h)$ denotes the number of cyclic factors of h which are of length k, for $1 \le k \le |X|$, then we have for the sum of the weights of the equivalence classes ω^1 of $\sim_1$:

3.8 $$\sum w_{\omega^1} = \frac{1}{|H|} \sum_{h \in H} \prod_{k=1}^{|X|} (z_1^k + \ldots + z_{|Y|}^k)^{a_k(h)} \ .$$

This means that the sum of the weights of the classes is obtained from the <u>cycle-index polynomial</u>

3.9 $$Z(H) = \frac{1}{|H|} \sum_{h \in H} \prod_{k=1}^{|X|} z_k^{a_k(h)}$$

by substituting $z_1^k+\ldots+z_{|Y|}^k$ for the indeterminate z_k. This is Pólya's famous result: the sum of the weights of the equivalence classes ω^1 of "$\sim_1$" is equal to

3.10 $$Z(H \mid \sum_1^{|Y|} z_i) \quad ,$$

i.e. it arises by substituting the sum $\sum z_i$ of the weights of the elements of Y in the cycle-index polynomial. Corresponding results for the equivalence relations "$\sim_2$" and "$\sim_3$" arise in the same way by an application of the weighted form of Burnside's lemma after having derived the number of fixed points for a general element of the group in question. A few remarks concerning this may be in order.

In order to get from $w^*: Y \to \mathbb{Q}[z_1,z_2,\ldots]$ a weight function $w: Y^X \to \mathbb{Q}[z_1,z_2,\ldots]$ by putting

$$w(\varphi) = \prod_X w^*(\varphi(x))$$

which is admissible, i.e. constant on the equivalence classes of "$\sim_3$" (and hence also of "$\sim_2$") we have to require that w^* is constant on the orbits of G on Y. The classes of "$\sim_3$" are just the orbits of the permutation representation δ of the wreath product

3.11 $$G \wr H = \{(f,h) \mid f: X \to G, h \in H\}$$

defined by

3.12 $$\delta: G \wr H \Rightarrow S_{Y^X}: (f,h) \longmapsto \binom{\varphi}{\psi} \quad ,$$

where $\psi \in Y^X$ is defined by

3.13 $$\psi(x) = f(x)\varphi(h^{-1}(x)) .$$

In order to get the number of functions $\varphi \in Y^X$ which remain fixed under this permutation $\delta((f,h))$ we may use the representation theory of wreath products (see (11)). It yields the following result. If again

$$h = \prod_{\nu} (j_\nu \dots h^{l_\nu - 1}(j_\nu))$$

is the cycle decomposition of h (which is uniquely determined if we require $j_\nu \leq h^s(j_\nu)$, for each s, and $j_\nu < j_{\nu+1}$), then we have the corresponding cycle products

3.14 $$g_\nu(f,h) = f(j_\nu)\, f(h^{-1}(j_\nu)) \dots f(h^{-l_\nu + 1}(j_\nu)) \in G$$

and in terms of these the number of fixed points of $\delta((f,h))$ satisfies

3.15 $$a_1(\delta((f,h))) = \prod_{\nu} a_1(g_\nu(f,h)) .$$

Hence, by Burnside's lemma, the number of classes of "$\sim_3$" is equal to

3.16 $$\frac{1}{|G|^{|X|}\,|H|} \sum_{(f,h)\in G\sim H} \prod_{\nu} a_1(g_\nu(f,h)).$$

For the sum of the weights w_{ω^3} of the classes ω^3 we obtain

3.17 $$\frac{1}{|G|^{|X|}\,|H|} \sum_{(f,h)\in G\sim H} \prod_{\nu} \sum_{g_\nu(f,h)(y) = y} w(y)^{l_\nu} .$$

This result, which is due to W. Lehmann (ref. (12),(13)), yields the corresponding results for "$\sim_2$" by just restricting attention to the following subgroup of $G\sim H$:

3.18 $$\{(f,h) \mid f\colon X \to G \text{ constant}, h \in H\} .$$

We obtain for the sum of the weights w_{ω^2} of the classes ω^2 of "$\sim_2$":

3.19 $$\frac{1}{|G|\,|H|} \sum_{(g,h) \in G \times H} \prod_{k=1}^{|X|} \Big(\sum_{g^k(y)=y} w(y)^k \Big)^{a_k(h)} .$$

Putting G = {1} we obtain Pólya's result 3.8 Hence both 2.10 (i) and (ii) (for properties which can be expressed by weight functions w which are obtained from mappings w which are constant on the orbits of G) are solved for each "$\sim_i$", i = 1,2,3.

2.10 (iii) is still far from a satisfactory solution. An algorithm which yields a complete system of representatives is available, it uses double-cosets. Using a big computer, it can be carried through for graphs with no multiple edges up to 8 points only without needing too much time. This sounds very poor, but we should keep in mind that there are already 12346 types of such graphs, and that $\binom{8}{2}$ = 28, so that it amounts to the evaluation of systems of double-cosets in S_{28}, which is quite a big group. Furthermore a program which works out a system of representatives of the classes directly is of no use either, since we are dealing with a set of 2^{28} mappings in this case. Hence I may conclude this section by asking for algorithms which give systems of representatives of the classes of "$\sim_2$" and "$\sim_3$".

4. A generalization

In the preceding section we saw how we can get a unified treatment for the count as well as for the count by weight of the classes of "$\sim_i$", $i = 1,2,3$. Hence one might suppose everything would be alright now as far as counting or counting by weight is concerned. But this feeling is false since besides the problems concerning the equivalence relations "$\sim_i$" which may be called Pólya enumeration theory, there is another branch of problems and results, which goes back to an in fact earlier paper, written by J. H. Redfield in 1927 (ref. (14)). It deals with so-called superpositions of graphs. Without giving a precise definition of this notion of superposition, let me just sketch the types of the superpositions of the following two graphs:

It can be shown that there are exactly 4 such types, namely

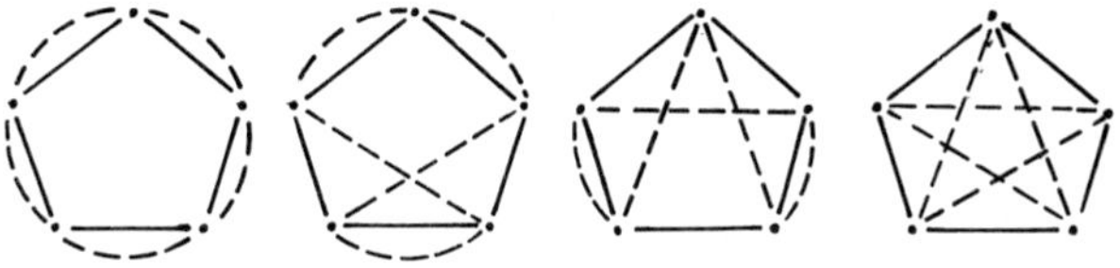

Redfield showed in his paper that the enumeration of the types of superpositions is a special case of the following enumeration problem. We consider the M of all the m × n-matrices $A = (a_{ik})$, which contain in each of their rows all the numbers $1,\dots,n$, so that each of these numbers occurs exactly one in each row of A.

On this set we introduce the following equivalence relation:

4.1 $$A \sim B \iff \exists\, \pi \in S_n \; (a_{ik} = b_{i,\pi^{-1}(k)}).$$

Hence A and B are called equivalent if and only if A arises from B by a permutation of the columns. The equivalence class of (a_{ik}) is denoted by

$$(a_{ik})_\sim .$$

If we are given m subgroups G_i of S_n, then we can consider the following permutation representation δ of $G_1 \times \ldots \times G_m$ on the set $M_\sim$ of equivalence classes:

4.2 $$\delta: (g_1,\ldots,g_m) \longmapsto \begin{pmatrix} (a_{ik})_\sim \\ (g_i(a_{ik}))_\sim \end{pmatrix} .$$

The main result of Redfield is that the number of orbits of $\delta[G_1 \times\ldots\times G_m]$ is obtained by applying a so-called <u>cap-operation</u> "$\cap$" to the cycle-indices of the groups G_i.

I would like now to introduce a more general problem which yields the result of the Pólya theory by just specializing. The aim was to give a unified treatment of both the Pólya and Redfield theory and on the other hand to make this whole theory more lucid in order to prepare it for further applications by making it easier to understand. I have the feeling that in particular the Redfield theory will have further applications which go far beyond the theory of counting isomers. In order to do this let me first show what we have to keep in mind if we want to generalize both the Pólya and the Redfield case. We notice first that a mapping $\psi: \{1,\ldots,m\} \to \{1,\ldots,m\}$ can be regarded as an $(m \times 1)$-matrix

$$\begin{pmatrix} \psi(1) \\ \vdots \\ \psi(m) \end{pmatrix} ,$$

so that the mapping φ with the values (of 2.9 (iii))
$\varphi(i) = f(i)\ \ \psi(h^{-1}(i))$, $f: \{1,\ldots,m\} \to G, h \in S_m$, is represented by

$$\begin{pmatrix} \vdots \\ \varphi(i) \\ \vdots \end{pmatrix} = \begin{pmatrix} \vdots \\ f(i)\psi(h^{-1}(i)) \\ \vdots \end{pmatrix} .$$

Hence besides permutations of the columns (which lead from M to M) and permutations acting on the entries of a row (cf. 4.2), we have to consider permutations of the rows. This gives the following idea. Instead of the set M introduced above we consider for a given $s \leq n$ the set

$$M^{<s>} ,$$

which consists of all the $(m \times s)$-matrices which contain in each one of the m rows s pairwise different elements of $\{1,\ldots,n\}$. On $M^{<s>}$ we have the same equivalence relation like on M (cf. 4.1) so that we obtain the set $M_{\sim}^{<s>}$ of equivalence classes of these $(m \times s)$-matrices. These equivalence classes are the orbits of the following permutation representation ε of the symmetric group S_s:

4.2
$$\varepsilon : \pi \longmapsto \begin{pmatrix} (a_{ik}) \\ \\ (a_{i,\pi^{-1}(k)}) \end{pmatrix} .$$

Besides this permutation representation we shall consider another one which we introduce next. In order to do this, we have to introduce the wreath product $(G_1,\ldots,G_r) \wr H$ of groups $G_1,\ldots,G_r$ with the subgroup H of S_m. which has the orbits $\omega_1,\ldots,\omega_r$:

$$(G_1,\ldots,G_r) \wr H = \{(g,h) \mid f: \{1,\ldots,m\} \to G_i, \forall i (i \in \omega_j \Rightarrow f(j) \in G_j)\} ,$$

together with the multiplication

$$(f,h)(f',h') = (ff'_h, hh') ,$$

where

$$f'_h(i) = f'(h^{-1}(i)) .$$

This group $(G_1,\dots,G_r)\sim H$ has the following permutation representation $\delta^{<S>}$ on the set $M^{<S>}$:

4.3 $$\delta^{<S>} : (f,h) \mapsto \begin{pmatrix} (a_{ik}) \\ \\ (f(i)\; a_{h^{-1}(i),k}) \end{pmatrix} .$$

We notice that by 4.2 and 4.3 we have for each $\sigma \in S_s$, each $(f,h) \in (G_1,\dots,G_r)\sim H$, and each $A \in M^{<S>}$:

4.4 $$\delta^{<S>}((f,h))\;\varepsilon(\sigma) = \varepsilon(\sigma)\delta^{<S>}((f,h)) ,$$

so that $\delta^{<S>}$ and ε satisfy the assumptions of the following lemma:

4.5 de Bruijn's lemma:

Let A and B denote two finite groups with permutation representations $\alpha : A \to S_M, \beta : B \to S_M$, on the same finite set $\mathbf{M}$ and such that for each triple (a,b,m) of $A \times B \times \mathbf{M}$ there exists an $b' \in B$ such that

$$\alpha(a)(\beta(b)(m)) = \beta(b')(\alpha(a)(m)),$$

then we have a representation $\tilde{\alpha}$ of A on the set $M_{\sim}$ of orbits of $\beta|B|$ as follows (denote by $\tilde{m}$ the orbit of m):

$$\tilde{\alpha} : a \mapsto \begin{pmatrix} \tilde{m} \\ \sim \\ \alpha(a)(m) \end{pmatrix} .$$

The number of orbits fixed under $\tilde{\alpha}(a)$ is

$$a_1(\tilde{\alpha}(a)) = \frac{1}{|B|} \sum_{b \in B} |\{m \mid \alpha(a)(m) = \beta(b)(m)\}| .$$

4.4 shows that we can apply 4.5 to $\delta^{<S>}$ and ε, so that we obtain a representation $\tilde{\delta}^{<S>}$ of A on $M_{\sim}^{<S>}$. Its character turns out to be as follows (ref. (10),(13)):

4.6

$$a_1(\widetilde{\delta^{<s>}((f,h))}) = \frac{1}{s!} \sum_{\sigma \in S_s} \prod_{\nu=1}^{c(h)} \prod_{k=1}^{s} \binom{a_k(g_\nu(f,h))}{a_k(\sigma^{-1}{}_\nu)} a_k(\sigma^{-1}{}_\nu)!\, k^{a_k(\sigma^{-1}{}_\nu)} .$$

A corollary is for example

4.7

$$a_1(\widetilde{\delta^{<m>}((f,1))}) = \frac{1}{m!} \sum_{\sigma \in S_m} \prod_{\nu} \prod_{k} \binom{a_k(f(\nu))}{a_k(\sigma)} a_k(\sigma)!\, k^{a_k(\sigma)}$$

$$= \begin{cases} (\prod_k a_k!\, k^{a_k})^{n-1}, & \text{if all the } f(\nu) \text{ are of the same cycle type } (a_1,\dots,a_n) \\ 0, & \text{otherwise.} \end{cases}$$

This corollary explains the result of Redfield, for his cap-operation between polynomials is defined as the linear extension of

$$\underbrace{(x_1^{b_1}\dots x_m^{b_m}) \cap \dots \cap (x_1^{c_1}\dots x_m^{c_m})}_{n \text{ monomials}} := \begin{cases} (\prod_k k^{b_k} b_k!)^{n-1}, & \text{if } b_1 = \dots = c_i \\ 0, & \text{otherwise.} \end{cases}$$

The results obtained above for the number of classes of "$\sim_i$", $i = 1,2,3$ come from the further corollary:

4.8

$$a_1(\delta^{<1>}((f,h))) = \prod_\nu a_1(g_\nu(f,h)).$$

REFERENCES

1. N. L. Biggs/E. K. Lloyd/ R. J. Wilson:
Graph Theory, 1736-1936.
Oxford University Press, 1976.

2. Crum Brown: On the theory of isomeric compounds.
Trans. Roy. Soc. Edinburgh 23 (1864), 707-719.

3. J. J. Sylvester: Chemistry and algebra.
Nature 17 (1877/78), 284.

4. A. Cayley: On the mathematical theory of isomers.
Philosophical Magazine (4) 47 (1874), 444-446.

5. A. C. Lunn/J. K. Senior: Isomerism and configuration.
J. Phys. Chem. 33 (1929), 1027-1079.

6. G. Pólya: Kombinatorische Anzahlbestimmungen für Gruppen, Graphen und chemische Berbindungen.
Acta Math. 68 (1937), 145-254.

7. F. Harary/E. Palmer: Graphical enumeration.
Academic Press, 1973.

8. A. Kerber: On Graphs and the Enumeration, Part I.
MATCH 1 (1975), 5-10.

9. A. Kerber: On Graphs and their Enumeration, Part II.
MATCH 2 (1976), 17-34.

10. A. Kerber/W. Lehmann: On Graphs and their Enumeration, Part III.
MATCH 3 (1977), 67-86.

11. A. Kerber: Representations of permutation groups I/II.
Lecture Notes in Math., Vol 240 and Vol. 495,
Springer Verlag 1971 and 1975.

12. W. Lehmann: Die Abzähltheorie von Redfield-Pólya-de Bruijn und die Darstellungstheorie endlicher Gruppen.
Diplomarbeit, Geißen 1973.

13. W. Lehmann: Ein vereinheitlichender Ansatz für die Redfield-Pólya-de- Bruijnsche Abzähltheorie,
Dissertation, Aachen, 1976.

14. J. H. Redfield: The theory of group reduced distributions,
Amer. J. Math. 49 (1927), 433-455.

Application of the Permutation Group to Stereoisomer Generation for Computer Assisted Structure Elucidation

James G. Nourse
Computer Science Dept.
Stanford University
Stanford Calif. 94305 USA

The purpose of this paper is to describe a recent effort to generate, classify, and enumerate the stereoisomers of all organic chemical structures consistent with a given empirical formula. The resulting theory is then converted into an algorithm which is programmed for a computer. The resulting program is used as an aid for computer assisted structure elucidation (1). The proper group theoretical formulation is the key to the solution of the problem and will be discussed here. Simply stated, the problem is to generate, classify and enumerate the possible stereoisomers for an organic chemical structure of defined constitution. (Constitution is the specification of the atoms in the molecule and their connectivity by bonds).

In order to formulate the desired symmetry group it is necessary to describe the constitution of a chemical structure with a graph in which the atoms correspond to the nodes and the bonds correspond to the edges. Multiple bonds correspond to multiple edges so the desired graph is actually a multigraph. All the nodes and edges are labelled uniquely (e.g. numbered consecutively). For the purposes here the symmetry of this graph will be described by a permutation group which includes all the node symmetries plus operations which interchange double edges (but not triple edges). This group is a semidirect product of the node symmetry group and the group which interchanges double edges and is called the Graph Symmetry Group.

The desired symmetry group is called the Configuration Symmetry Group (CSG) and is defined to be the Graph Symmetry Group represented by its action on the configurations of all trivalent and tetravalent atoms in the chemical structure. The configurations at these atoms are determined by the numbers of the substituents at each atom. Thus two enantiomeric configurations can be defined for any tetravalent atom or nonplanar trivalent atom by simply giving all the substituents different numbers, even for atoms which do not exist in enantiomeric forms.

The representation of the Graph Symmetry Group on the configurations of the atoms is constructed by determining the effect of the graph permutations on these configurations. This is done by defining an ordering of the four substituents on all the atoms. This ordering is most conveniently the ordering induced by the numbering of the atoms. If a graph permutation fixes an atom it will either leave the configuration of that atom unchanged (an even permutation of the substituents) or it will invert its configuration (an odd permutation of the substituents). This information is added to the permutation. If a permutation takes one atom to another then a correspondence of the substituents must be defined. This correspondence is determined by the ordering of the substituents. Thus if atom A goes to atom B, then there is a mapping of the four substituents on atom A to the four substituents on atom B.

$$(a_1, a_2, a_3, a_4) \text{ ---- > } (b_1, b_2, b_3, b_4)$$

If the permutation of the four numbered substituents defined by this mapping is even, then the configuration of A remains unchanged after it is mapped to B. If this permutation is odd, then the configuration of A inverts as it is mapped to B. This procedure is performed for all the permutations in the graph symmetry group and for all the atoms with three (nonplanar) or four substituents. The resulting group is called the Configuration Symmetry Group. Intuitively, this group is an invariance group for a stereoisomer in which the relative configurations of constitutionally identically substituted atoms is taken into account. See ref. (2) for a pictorial example.

The resultant operations in the CSG are actually permutation inversions where the inversions are of individual atom configurations. The group of all such permutation inversions is the wreath product $S_n[S_2]$ for n atoms (3). The permutation inversions in the CSG act on the 2^n possible stereoisomers to yield equivalence classes which correspond to the symmetrically distinct stereoisomers. Each possible stereoisomer can be represented by an n-tuple of the configurations of the atoms. Thus for example consider tetramethylcyclobutane 1, which has four carbon atoms which are capable of existing in two distinct configurations. For simplicity the methyl carbons are not considered.

There are 2^4 or 16 potential stereoisomers of this structure. Each of these stereoisomers can be represented as a 4-tuple such as [+-+-] which means that atom 1 is in the "+" configuration (based on the atom numbering), atom 2 is in the "-" configuration etc. These 4-tuples correspond to chiral graphs (4) which are graphs augmented with "+" or "-" parity labels at the trivalent and tetravalent nodes. The configuration symmetry group for this structure is derived as described above and is given in the table. The action of these permutation inversions on the 4-tuples is (chiral graphs) as follows: A permutation inversion such as (12'34') is read: 1 goes to 2 and inverts configuration, 2 goes to 3, 3 goes to 4 and inverts configuration, and 4 goes to 1. This acts on the 4-tuple [-+++] as:

(12'34') [-+++] -----> [+++-]

Doing this for all the permutation inversions in the CSG on all the 4-tuples has the effect of collecting the 16 4-tuples into 4 equivalence classes which correspond to the 4 possible stereoisomers of tetramethylcyclobutane. These are shown in the table and the structures.

	2a	2b	2c	2d
(1)(2)(3)(4)	[++++]	[-+++]	[--++]	[-++-]
(1')(24)(3')	[-+-+]	[++-+]	[++--]	[+--+]
(12)(34)	[++++]	[+-++]	[--++]	[+--+]
(12'34')	[+-+-]	[+++-]	[++--]	[-++-]
(13)(2')(4')	[+-+-]	[+---]	[++--]	[+--+]
(1'3)(2'4')	[----]	[--+-]	[--++]	[-++-]
(1'43'2)	[-+-+]	[-+--]	[++--]	[-++-]
(1'4')(2'3')	[----]	[---+]	[--++]	[+--+]

1

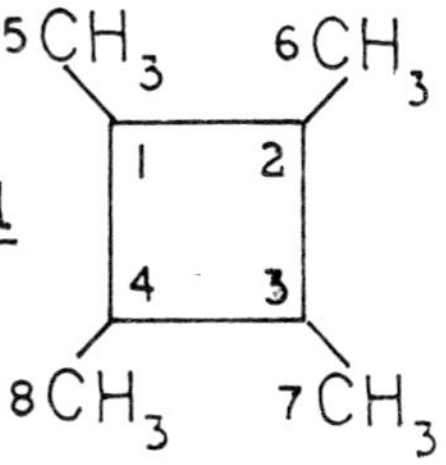

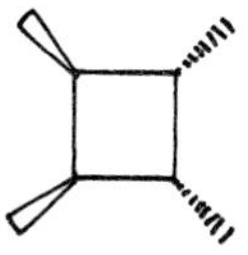

2a

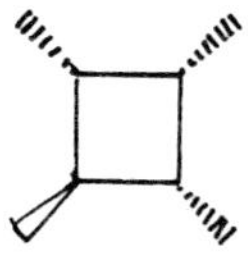

2b

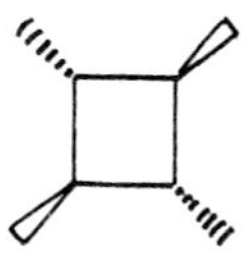

2c

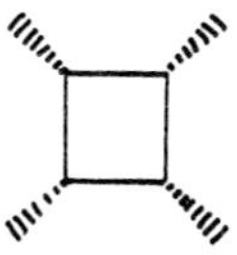

2d

This method of forming equivalence classes of the possible stereoisomers can be used in an algorithm which generates all the distinct stereoisomers of a chemical structure of defined constitution. This algorithm has been implemented as a computer program used for structure elucidation (5).

The specification of the equivalence class of the CSG to which a given stereoisomer belongs provides a means of classification of stereoisomers. When combined with some means of uniquely (canonically) numbering the atoms of the chemical structure, a unique (canonical) name for the structure results. Each stereoisomer is represented by the numbered graph describing its constitution with parity labels at each nonplanar trivalent and tetravalent atom. For each equivalence class of the CSG a representative member is chosen which is in some sense unique. This can be the "lowest" member of the equivalence class where "+" is considered to be lower than "-" and the atoms are ordered by their node numbers (6). This specification of the configuration of a stereoisomer is independent of geometry since only a graph (usually represented tabularly) augmented with parity labels (the chiral graph) is required to uniquely represent the stereoisomer. While these parity labels ultimately refer to geometric configurations, there is no use made of any geometric property when these parity labels are determined. Only the connectivity of the numbered graph and the parity of permutations is required. Thus the configuration (by this specification) of a stereoisomer can be separated from any geometric property. This may lead to a separation of the notions of configuration (nongeometric) and conformation (geometric) at least for typical organic structures.

This classification of stereoisomers simultaneously considers global and local properties of the structure which facilitates providing a unique name. The local property is the configuration of the atoms based on the numbering of the adjacent substituents. The global property is the overall symmetry described by the configuration symmetry group. The final specification gives just the configuration designations which were computed by using the CSG. The CSG need not be part of the designation. A specification of the configuration of a stereoisomer which relies only on local properties such as the current R/S naming system becomes very difficult when the structure has a great deal of overall symmetry (7). A specification which relies only on a global property such as the symmetry group of the stereoisomer would not provide a unique specification since most chemical structures have little or no overall symmetry.

Using the Configuration Symmetry Group, a single counting formula can be derived which gives the number of distinct stereoisomers for a structure of defined constitution. This problem dates back to van't Hoff and no general solution has been heretofore available. The desired counting formula is one which requires only the symmetry group of the structure in analogy with the Polya enumeration theorem (8).

Combinatorial problems of this kind are generally formulated as mapping problems in which the number of equivalence classes of mappings under the action of some symmetry group is determined.

In this case the tri- and tetravalent atoms are mapped to the two possible configuration "+" or "-" based on the atom numberings. The problem is now to count the number of equivalence classes of these mappings formed by the action of the CSG. It has already been pointed out that the CSG is a subgroup of the wreath product group $S_n[S_2]$. The permutation representation of this wreath product is an example of an exponentiation group as described by Kerber (9). For this problem, the exponentiation group is the wreath product of the group of all identical atom permutations (S_n in this case for n identical atoms) around the group which interchanges the two configurations (S_2 in this case). The exponentiation group therefore includes all the possible domain and range symmetries possible for the above mapping of atoms to configurations. The CSG is a subgroup of this exponentiation group. Kerber has given the general counting formula for the exponentiation group (9) so that counting the equivalence classes of the CSG and hence the possible stereoisomers requires the restriction of this formula to the CSG subgroup.

The counting formula will be illustrated with the tetramethylcyclobutane example, 2a-d. The computation is summarized by:

	(1)(2)(3)(4)	(12'34') (1'43'2)	(1'3')(2'4')	(1'4')(2'3') (12)(34)	(1')(24)(3') (13)(2')(4')
order	1	2	1	2	1
orbits	4	1	2	2	3
orbits with odd inv.	0	0	0	0	2
contribution	16	4	4	8	0

Each conjugacy class in the group contributes a term to the total equal to its order times 2^t where t is the number of orbits in a permutation in that conjugacy class. The computation at this stage resembles the familiar Polya counting method. The difference from the Polya method is that any conjugacy class which includes an orbit with an odd number of inversions will contribute zero to the total. In the example there is only one conjugacy class with orbits including an odd number of inversions. The total number of stereoisomers is then the total of the contributions from each conjugacy class divided by the order of the group. This gives 32/8 = 4 stereoisomers for tetramethylcyclobutane and these are indicated by structures 2a-d.

The stereoisomer counting formula is:

$$\frac{1}{g} \sum_{i=1}^{c} h_i \cdot \prod_{j=1}^{p} 2^{n_j} \cdot \prod_{k=0}^{n_j} (n_{ijk} + 1) \bmod 2$$

g = order of CSG

c = number of conjugacy classes in CSG

h_i = order of the ith conjugacy class

p = number of atoms for which configurations are counted

n_j = number of orbits of length j

n_{ijk} = number of inversions in the kth cycle of length j in the ith conjugacy class, $n_{ij0} = 0$

As a second example, the computation of the number of stereoisomers of 3 is summarized below. The CSG for 3 has order 72. Four conjugacy classes have cycles with an odd number of inversions and contribute zero to the total. There are 216/72 = 3 stereoisomers.

class	1^6	$1^4 2$	$1^3 3$	$1^2 22$	123	6	24	2^3	3^2
order	1	6	4	9	12	12	18	6	4
orbits	6	5	4	4	3	1	2	3	2
contr.	64	0	64	0	0	24	0	48	16

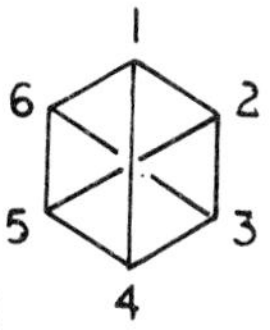

3

It can be easily proved that a permutation inversion with an odd number of inversions in an orbit will contribute zero to the counting total. By Burnside's lemma the contribution of a group element to the counting total will be equal to the number of objects fixed by that group element. In this cases these objects are the 2^n possible stereoisomers. To show that an orbit with an odd number of inversions cannot fix an n-tuple, start at the beginning of the orbit. The permutation takes this first atom to another atom. If an inversion must be done here, the two atoms must have opposite configurations if this n-tuple is to be fixed by the permutation inversion. This second atom is taken to a third atom by the permutation. If an inversion is to be done here, then the second and third atoms must have opposite configurations and therefore the first and third must have the same configuration. This continues until the first atom is reached again. If there are an odd number of inversions in the orbit, the only way an n-tuple can be fixed is if the first atom has the opposite configuration of itself which is impossible. Hence, orbits with an odd number of inversions cannot fix any n-tuple.

This combinatorial result does not depend on the particular choice of numbering of the atoms in the chemical structure even though the exact form of the permutation inversions in the Configuration Symmetry Group do depend on this numbering. Renumbering the atoms corresponds to the operation of conjugation by an element of the group $S_n [S_2]$. The permutation which effects the numbering change is an element of this group. A renumbering will change the CSG into a conjugate CSG within the group $S_n [S_2]$. The key properties of the elements of this group for the counting formula are the orbit structures of the permutation inversions and the number of orbits with an odd number of inversions. Both of these properties are invariant to conjugation. Thus if a permutation inversion includes an orbit of length 4 with 3 inversions, it will go to an orbit of length 4 with either 1 or 3 inversions by the operation of conjugation (renumbering the atoms) and hence will make the same contribution to the counting total.

A limitation of this method of specifying and counting stereoisomers is that only the connectivity of the structure with parity designations is given and no account is taken of conformation (e.g. rotation around single bonds) or topology in the sense of a catenane (interlocked rings). Stated more exactly, structures which can be interconverted by rotations around single bonds or can be interconverted by bonds passing through bonds, are given the same configuration specification. The former case includes substituted biphenyls and the later case includes catenanes (interlocked rings).

Financial support from the National Institutes of Health (grant 2R24 RR 00612-08) is gratefully acknowledged.

References

1. (a) L. M. Masinter, N. S. Sridharan, J. Lederberg, and D. H. Smith, J. Amer. Chem. Soc., 96, 7702 (1974).

   (b) R. E. Carhart, D. H. Smith, H. Brown, and C. Djerassi, J. Amer. Chem. Soc., 97, 5755 (1975).

2. J. G. Nourse, J. Amer. Chem. Soc., 97, 4594 (1975).

3. C. A. Mead, Top. Curr. Chem., 49, 1 (1974).

4. R. W. Robinson, F. Harary, and A. T. Balaban, Tetrahedron, 32, 355 (1976).

5. J. G. Nourse, R. E. Carhart, D. H. Smith, and C. Djerassi, J. Amer. Chem. Soc., submitted.

6. W. T. Wipke and T. M. Dyott, J. Amer. Chem. Soc., 96, 4834 (1974).

7. R. S. Cahn, C. K. Ingold, and V. Prelog, Angew. Chem. Int. Ed. Engl., 5, 385 (1966). See footnote on p. 397. We are merely pointing out a difficulty in designing a system of nomenclature, not proposing a new system.

8. (a) G. Polya, Acta Math., 68, 145 (1937).

   (b) B. A. Kennedy, D. A. McQuarrie, and C. H. Brubaker, Jr., Inorg. Chem., 3, 265 (1964).

9. (a) A. Kerber, "Representations of Permutation Groups. II", Springer-Verlag, New York, 1975.

   (b) A. Kerber, Discrete Math., 13, 13 (1975).

Applications of the Permutation Group in Dynamic Stereochemistry

James G. Nourse
Computer Science Dept.
Stanford University
Stanford, Calif. 94305 USA

A typical problem in dynamic stereochemistry is to determine the mechanism of the rearrangement of a chemical structure such as 1.

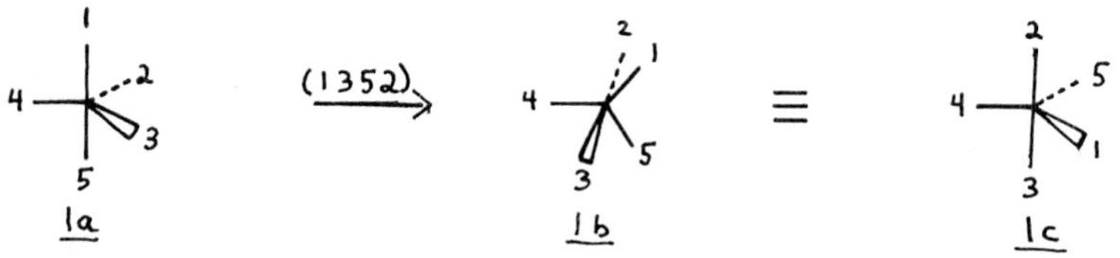

The ligands may be the same or different and the rearrangement may be skeletally degenerate or nondegenerate. Many of the problems already studied have involved degenerate rearrangements (those which interconvert chemical structures of the same geometry apart from any difference in the ligands) and only these will be considered here. Two groups and the mapping (homomorphism) between them are crucial to the proper description of problems of this kind. The first of these is the symmetry group of the chemical structure's skeleton (i.e., without ligands) and is designated P. This may be a point group, the rotation subgroup of a point group, or a nonrigid symmetry group (1). Generally, this group can be represented as a permutation group on the ligands or on the skeletal sites however, in the most general case, this group is best represented as a permutation-inversion group (1). In particular, when treating problems involving the degenerate rearrangement of chemical structures with chiral skeletons, the proper description of overall changes requires the permutation-inversion designation. For example, to describe the rearrangement of the chiral (C_3 symmetry) propeller molecule shown:

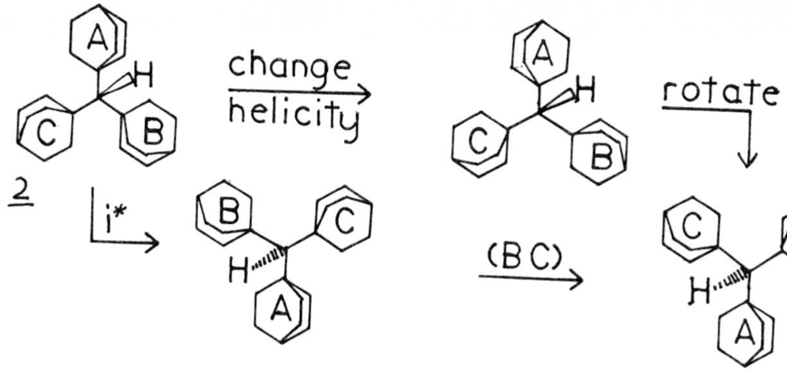

in which the configuration of the central carbon does not change but in which the helicity of the propeller (as defined by the dihedral angle between the phenyl rings and the central carbon-hydrogen bond) does change, the proper description is the permutation shown coupled with overall coordinate inversion. Coordinate inversion changes both the configuration of the central carbon (as determined by the labelling of the rings) and the helicity. Ring permutation changes only the configuration. Separately these operations are not feasible by the Longuet-Higgins criteria (1), however, the combined operation changes only the helicity and is feasible for these systems.

The second group is the group which includes all possible permutations of either the identical atoms in the molecule or the possible ligand sites on the molecule. The distinction between the two is important in many cases (2). This group is designated, G, and is usually one of the symmetric groups or a direct product of symmetric groups. However, in many cases only a subgroup is needed because of a feasibility condition (1). That is, only certain overall changes are considered to be feasible under the experimental conditions at hand. This group is also best represented as a permutation inversion group although a permutation representation is often adequate.

The homomorphism between these two groups is just the injective mapping of P into G:

$$P \dashrightarrow G$$

This necessarily preserves the permutation (inversion) representation. For example, consider the two structures 1 and 2. The relevant groups are:

Example	P	G
1	D_{3h}	S_5
2	C_3	$S_3[S_2] \times C_i$

For the trigonal-bipyramidal structure 1, the point group is D_{3h} and the rotation subgroup is D_3. The permutation group G is the symmetric group S_5 which exchanges all 5 sites in all possible ways. Because the skeleton is not chiral, the permutation-inversion designations are not necessary. For the triphenylmethane structure 2, the point group is C_3 as the skeleton is chiral. The permutation inversion group is the direct product of the wreath product $S_3[S_2]$ and the overall coordinate inversion C_i. However, if inversion of configuration of the central carbon is not considered feasible, then the group G is a subgroup which includes only those operations which do not invert the configuration of the central carbon as discussed above. Wreath products are very common in problems of this kind since they represent the feasible changes which correspond to exchanges within exchanges. Given this formulation of the problem a number of associations of group theoretical structures and chemical concepts can be made. Isomers of the chemical structure with all the sites substituted differently correspond to the cosets gR of the rotation subgroup $R < P$ where g is in G. If some of the ligands are identical, then the isomers correspond to the double cosets (3):

$$L / P \backslash R$$

where L is the symmetry group of the identical ligands and is usually a direct product of symmetric groups (4).

The possible modes of isomerization for a structure with all ligands identical correspond to the double cosets:

$$R / P \backslash R$$

or to unions of these double cosets (5). More generally these modes correspond to bilateral classes (6). Each mode is a collection of permutations which are made equivalent by the symmetry of the skeleton. Each permutation represents the overall change caused by a degenerate isomerization which can occur by a variety of physical mechanisms. In the well studied trigonal bipyramid example, there are six double cosets : $D_3 / S_5 \backslash D_3$ which correspond to the six modes or permutationally distinct isomerizations (5). One of these modes includes the permutation corresponding to the familiar Berry (7) pseudorotation indicated by the sequence of structures 1a-c.

If the ligands are not all identical then modes are defined by considering the ligand symmetry as well as the site symmetry and are called generalized stereoisomerization modes (2).

Another key group theoretical construction is the lattice of subgroups between P and G (8). Here P will be the rotation subgroup of the point group. Properties of this lattice and the subgroups involved can be used in at least three ways.

First, chains of subgroups between P and G can be used to provide descriptors for the possible isomers which correspond to the cosets gP in G. In most problems of interest there is a very large number of these isomers and a means of keeping track of them is important. This also provides a general method of classification of permutation isomers (9). Consider such a chain of subgroups between P and G:

$$P \dashrightarrow H_1 \dashrightarrow H_2 \dashrightarrow H_3 \dashrightarrow \cdots\cdots \dashrightarrow H_n \dashrightarrow G$$

Starting from the largest group G, the cosets of H_n in G collect the isomers into i_{nG} classes where i_{nG} is the index of H_n in G. Each class contains i_{Pn} isomers where i_{Pn} is the index of P in H_n. All the isomers in each coset gH_n receive the same descriptor while each coset gH_n receives a different descriptor. This process is repeated for the cosets of H_{n-1} in H_n etc. until P is reached. At each stage the isomers receive a new descriptor. Finally, there will be as many descriptors as there are subgroups in the chain. These are not absolute designations and depend on the choice of a reference isomer of some kind. Once this reference isomer is chosen and has its descriptors assigned, all the other isomers will have unique descriptors. This idea has been applied to the problem of giving descriptors to the 192 possible isomers of tetraphenylmethane (8). It is possible to choose any of the possible chains of subgroups and an optimal choice is not apparent (8). One choice with a kind of uniqueness property is a composition series which is a chain of maximal normal subgroups (10), each maximal in the group directly above it. By the Jordan-Holder theorem, all composition series in a group are equivalent although the actual chain of groups can vary.

A second use of the subgroup lattice concerns the concept of "residual stereoisomerism". It has been observed that for propeller-like molecules such as 2 with each ortho position distinguished, the preferred mode of rearrangement does not interconvert all the possible isomers and two distinct isomers are observed (11). In group theoretic terms, the double coset which corresponds to the energetically preferred mode of rearrangement when multiplied exhaustively times itself generates only a subgroup of the maximal permutation inversion group G. For a given problem with point group P and maximal group G, the possibilities for residual stereoisomerism correspond to the subgroups on the lattice between G and P. Each of these subgroups is composed of intact double cosets of G in P and each of these double cosets can correspond to a possible rearrangement mode. The actual preferred mode and hence the number of residual stereoisomers depends on the energetics of the problem and is outside the realm of the permutation group description.

The third use of the subgroup lattice involves the determination of the effect of combining two or more experiments to determine a preferred rearrangement mode. Here, use is made of the lattice operation of intersection. On the subgroup lattice the intersection of two groups is the largest subgroup common to both of them. Consider the subgroups:

$$H_1 \dashleftarrow P \dashrightarrow H_2$$

i.e., the intersection of subgroups H_1 and H_2 is P. It is possible that an experiment can be designed which has as its effective symmetry group the subgroup H_1. Thus the results of the experiment are invariant to operation in the group H_1 rather than just the smaller group P. Certain NMR experiments and experiments involving structures with identical substituents can have effective symmetry groups larger than P (8). Performing such an experiment would determine a rearrangement mode corresponding to a double coset of H_1 in G. This double coset is a collection of the double cosets of P in G since P is a subgroup of H_1. Doing another experiment with effective symmetry group H_2 would likewise determine a mode corresponding to a double coset of H_2 in G which would be a different collection of the double cosets of P in G. Now assuming the actual favored mode to be the same in both experiments, this mode must be common to the collections of modes determined for the two experiments. That is, the actual mode must be in the intersection of the two collections. The overall effect is that combining the two experiments with effective symmetry groups H_1 and H_2 has the effect of doing the one experiment with effective symmetry group P. More precisely, the upper limit of the information available by doing the two experiments with effective symmetry groups H_1 and H_2 is the information available by doing the experiment with effective symmetry group P which is the intersection of H_1 and H_2. The qualifier is necessary since the intersection of a double coset of H_1 with a double coset of H_2 need not give just one double coset of P. An example of the use of this idea has been given (7). The idea is useful only if the two experiments together are simpler to perform than the one which corresponds to their intersection.

In the usual sense the group G is generally not considered to be a symmetry group of the chemical structure in question except occasionally as a nonrigid symmetry group (1). However, G is a symmetry group for the potential energy surface which corresponds to the set of atoms which make up the structure since G includes permutations of identical atoms. The potential energy surface in question starts out with 3n+1 coordinates (3 for each of the n atoms and 1 energy coordinate). This is reduced by the identification of all the points on the surface which are related by translation or rotation in space. Each point on this surface represents a chemical structure in all possible orientations. For structures with nontrivial rotation symmetry, this process identifies structures related by the permutations of the rotation group, R. Hence the cosets gR are represented by single points. Starting from any of these points it is possible to find paths to another isomer (coset) which correspond to one of the possible modes of rearrangement. For example consider the Berry pseudorotation of a trigonal bipyramid structure:

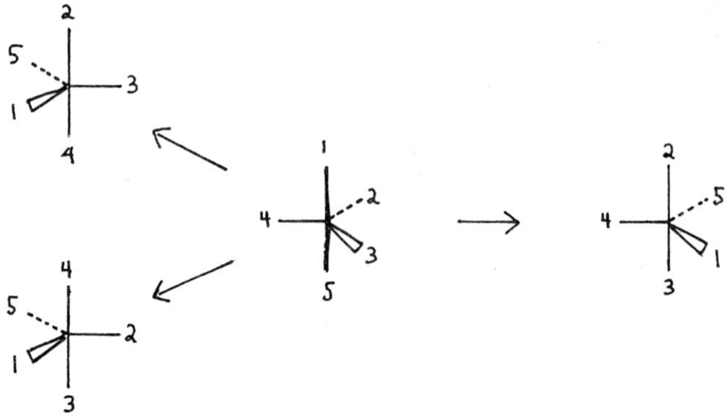

There are three possible symmetrically related Berry processes which have the indicated overall permutational change. Each of these permutations is in the same double coset or mode. Now the group G (S_5 in this case) acts on the surface (and the cosets) as a symmetry group relating all the isomers (20 cosets in this case). Similarly, this group relates all the symmetrically equivalent paths which correspond to the Berry process. This set of isomers and paths can be symbolized by a graph in which the nodes are the isomers and the edges are the paths. These interconversion graphs have seen wide application in problems in dynamic stereochemistry (12). The incidence matrices for these graphs are the double coset matrices discovered by Frame (13). These matrices commute with the permutation representation of G and therefore have G as a symmetry group. Another interesting symmetry property of the paths connecting the isomers on the potential energy surface follows from an intrinsic property of double cosets. If there is a path from isomer a to isomer b with overall change the permutation p, then there will be a symmetrically equivalent path from b to a if p^{-1} is in the same double coset as p. Stated differently, the double coset which includes p must be selfinverse (3). There is an overall symmetry of the potential energy surface which exchanges isomers a and b (14).

Further use can be made of the idea of group homomorphism to describe the observed similarity between seemingly diverse chemical systems termed stereochemical correspondence (15). For two chemical systems characterized by groups P_1, G_1, P_2, G_2 and the homomorphisms between P_1 and G_1 and between P_2 and G_2, there is a correspondence from chemical system 1 to chemical system 2 if there is a diagram of homomorphisms of the groups:

$$\begin{array}{ccc} P_1 & \longrightarrow & P_2 \\ \downarrow & & \downarrow \\ G_1 & \longrightarrow & G_2 \end{array}$$

that commutes. That is, composing the homomorphisms around the diagram in either direction yields the same overall homomorphism. This condition assures a correspondence between cosets (isomers) and double cosets (modes). Stronger conditions insure a correspondence between the subgroup lattices (15c). Such a correspondence can be thought of as a description of the similarity between regions of the potential energy surfaces describing the two chemical systems. This similarity is in the connectivity of the surface and not necessarily in the energetics. However, more closely related systems for which there is stereochemical correspondence have been observed to be similar energetically (11). The observation of such a correspondence aids in the design of experiments, particularly if one of the chemical systems has already been studied. This is also an interesting relationship between chemical systems in that it can be described algebraically in a very compact way (15c). This yields a new means of classifying chemical systems based on an intrinsic algebraic property of these systems and resembles the methods used to classify various mathematical structures by their intrinsic algebraic properties (16). A number of other examples of stereochemical correspondence have been given (15). A summary of the correspondence of some concepts in dynamic stereochemistry with their permutation group theoretical structures is given below.

Chemical Concept	Permutation Group Structure
maximally substituted isomers	cosets, gR
isomers with some identical substituents	double cosets, LgR
isomerization modes	(unions of) double cosets, RgR bilateral classes
residual stereoisomers (closed system of interconverting isomers)	subgroup between R and G
interconversion graphs (topological representations)	(unions of) double coset matrices
isomer descriptors	subgroup chains
combined experiments with different effective symmetry groups	subgroup lattice intersection

Financial Support from the National Institutes of Health (grant 2R24 RR 00612-08) is gratefully acknowledged.

References

1. H. C. Longuet-Higgins, Mol. Phys., 6, 445 (1963).
2. J. G. Nourse, J. Amer. Chem. Soc., 99, 2063 (1977).
3. J. S. Frame, Bull. Amer. Math. Soc., 47, 458 (1941).
4. E. Ruch, W. Hasselbarth, and B. Richter, Theor. Chim. Acta., 19 288 (1970).
5. (a) M. Gielen and N. Vanlautem, Bull. Soc. Chim. Belg., 79, 679 (1970).
   (b) P. Meakin, E. L. Muetterties, F. N. Tebbe, and J. P. Jesson, J. Amer. Chem. Soc., 93, 4701 (1971).
   (c) W. G. Klemperer, J. Chem. Phys., 56, 5478 (1972).
   (d) W. Hasselbarth and E. Ruch, Theor. Chim. Acta, 29, 259 (1973).
   (e) D. J. Klein and A. H. Cowley, J. Amer. Soc., 97, 1633 (1975).
6. W. Hasselbarth, E. Ruch, D. J. Klein, and T. H. Seligman, in "Group Theoretical Methods in Physics, R. T. Sharp and B. Kolman, eds., Academic, New York, 1977, p. 617.
7. R. S. Berry, J. Chem. Phys., 32, 923 (1960).
8. J. G. Nourse and K. Mislow, J. Amer. Chem. Soc., 97, 4571 (1975).
9. Related ideas of descriptors and "stereochemical quantum numbers" have been discussed. See E. Ruch and I. Ugi, Top. Stereochem., 4, 99 (1969).
10. J. J. Rotman, "The Theory of Groups, An Introduction", Allyn and Bacon, Boston, 1965, chap. 6.
11. K. Mislow, Acc. Chem. Res., 9, 26 (1976).
12. (a) M. Gielen, R. Willem, and J. Brocas, Bull. Chim. Soc. Belg., 82, 617, (1973).
    (b) K. Mislow, Acc. Chem. Res., 3, 321, (1970).
13. J. S. Frame, Bull. Amer. Math. Soc., 49, 81 (1943); 54, 740 (1948).
14. J. G. Nourse, to appear.
15. (a) D. Gust, P. Finocchiaro, and K. Mislow, Proc. Natl. Acad. Sci. U.S., 70, 3445 (1973).
    (b) K. Mislow, D. Gust, P. Finocchiaro, and R. J. Boettcher, Fortschr. Chem. Forsch. 47, 1-28.
    (c) J. G. Nourse, Proc. Natl. Acad. Sci. U.S., 72, 2385 (1975).
16. S. MacLane, "Categories for the Working Mathematician", Springer-Verlag, New York, 1971.

The Spin Double Groups of Molecular Symmetry Groups

by

P.R. Bunker
Herzberg Institute of Astrophysics
National Research Council of Canada
Ottawa, Ontario, Canada K1A 0R6

Introduction

The energy levels of molecules are labelled, using quantum numbers and irreducible representations, in order to assist in the understanding of the effects of both intramolecular interactions and externally applied perturbations. The irreducible representations of the three-dimensional rotation group $\underset{\sim}{K}$ and of the molecular symmetry (MS) group are particularly useful labels. In this paper I will discuss as examples the determination of these labels for the molecules CH_3F, CH_3F^+ and $CH_3BF_2^+$; the two ionic molecules have an odd number of electrons and the electron spin double groups are required for them.

Molecular Wavefunctions

To determine the irreducible representation labels for the energy levels of a molecule we must first determine appropriate approximate wavefunctions. The representations generated by these wavefunctions give us the required labels. By 'appropriate' approximate wavefunctions we mean wavefunctions obtained as eigenfunctions of an approximate Hamiltonian whose eigenvalues have the same pattern (semi-quantitatively) as the observed energy levels; in other words the approximations made must be reasonable.

We can usually consider as appropriate the wavefunctions obtained after making the following series of approximations:

(i) The Born-Oppenheimer approximation,

(ii) The LCAOMO approximation to the electronic wavefunction,

(iii) The rigid rotor-harmonic oscillator approximation to the rotation-vibration wavefunction, and

(iv) The approximation of neglecting the magnetic coupling of the spins to the overall rotational motion.

In these circumstances the approximate wavefunctions are separable and can be written as

$$\Phi_{nspin} \; \Phi_{rot} \; \Phi_{vib} \; \Phi_{elec} \; \Phi_{espin}. \tag{1}$$

We can classify each of the component parts by symmetry and the overall symmetry will be the product of these individual symmetries. We are often only concerned with the rovibronic symmetry, i.e. that of $\Phi_{rve} = \Phi_{rot}\Phi_{vib}\Phi_{elec}$.

For a symmetric top molecule such as CH_3F the rigid rotor wavefunctions are

$$S_{J,k,m}(\theta,\phi,\chi) = \Theta_{J,k,m}(\theta)e^{im\phi}e^{ik\chi} \quad , \tag{2}$$

where the Euler angles (θ,ϕ,χ) are shown at the top of Fig. 1. To determine the transformation properties of this function under the effect of a symmetry operation we need to know that

$$\Theta_{J,k,m}(\pi-\theta) = (-1)^{J-m} \; \Theta_{J,-k,m}(\theta) \; . \tag{3}$$

$J(J+1)\hbar^2$ is the total angular momentum, $k\hbar$ is the component along the molecule fixed z axis, and $m\hbar$ is the component along the space fixed ζ axis.

In this approximation Φ_{vib} is the product of harmonic oscillator functions (i.e. hermite polynomials) in the normal coordinates, and Φ_{elec} is the product of molecular orbital functions each of which is written as a linear combination of atomic orbital functions. The normal coordinates are linear combinations of the xyz nuclear displacements away from equilibrium, and their transformation properties can be determined. The vibrational ground state wavefunction is always totally symmetric in the molecular symmetry group. The symmetry of Φ_{elec} is obtained as the product of the symmetry of the molecular orbitals of which it is composed,and the symmetries of the

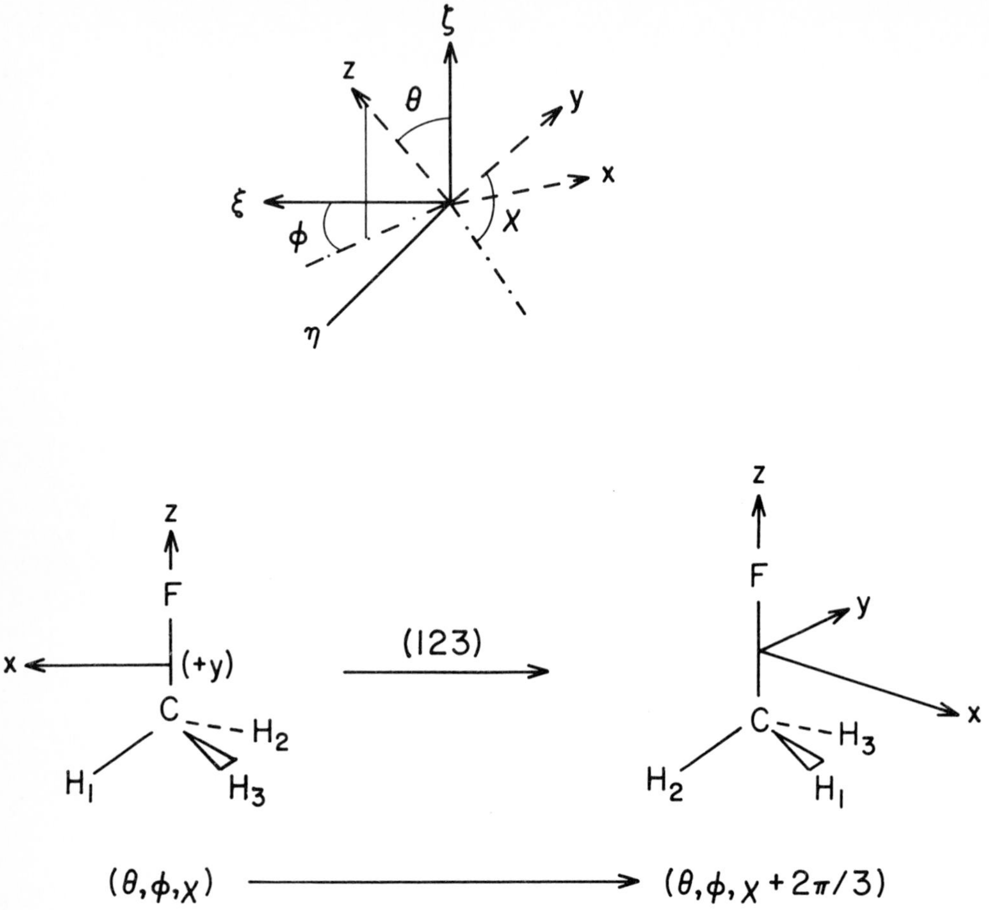

Fig. 1. The Euler angles for CH_3F

molecular orbitals can be determined from the transformation properties of the atomic orbitals.

$\underline{CH_3F}$

Labelling the protons 1, 2 and 3 the MS group of CH_3F is {E,(123),(132),(12)*,(23)*,(13)*}, which we can call $\underset{\sim}{C}_{3v}(M)$. The character table is given in Table I. To label the lowest energy levels (the rotational energy levels) we must determine the representations generated by the rotational wavefunctions $S_{J,k,m}(\theta,\phi,\chi)$. This is easily accomplished once we have determined the transformation properties of the Euler angles (see, for example, Fig. 1).

To discuss the transformation properties of the Euler angles it is convenient to introduce two rotation operations, R_z^β and R_α^π, which are defined as follows:

$$R_z^\beta(\theta,\phi,\chi) = (\theta,\phi,\chi+\beta) \qquad (4a)$$

$$R_\alpha^\pi(\theta,\phi,\chi) = (\pi-\theta,\phi+\pi,2\pi-2\alpha-\chi) \qquad (4b)$$

We can identify a rotation operation with each element of the MS group in order to specify the Euler angle change caused, and for CH_3F the identifications are as follows:

E	(123)	(132)	(12)*	(23)*	(13)*
R^0	$R_z^{2\pi/3}$	$R_z^{4\pi/3}$	$R_{\pi/6}^\pi$	$R_{\pi/2}^\pi$	$R_{5\pi/6}^\pi$

Using these results we can classify the $S_{J,k,m}(\theta,\phi,\chi)$ in $\underset{\sim}{C}_{3v}(M)$ to obtain the species of the rotational states. The vibrational and electronic wavefunctions are totally symmetric (A_1) for the ground state so the species of the rovibronic states are as given in Fig. 2. The angular momentum quantum number label J can be considered as the irreducible representation label $D^{(J)}$ from the group $\underset{\sim}{K}$.

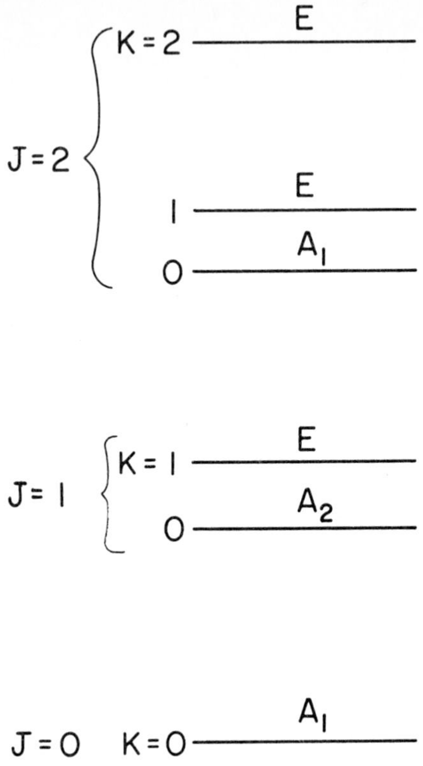

Fig. 2. The energy levels of CH_3F in a $^1A_1$ vibronic state.

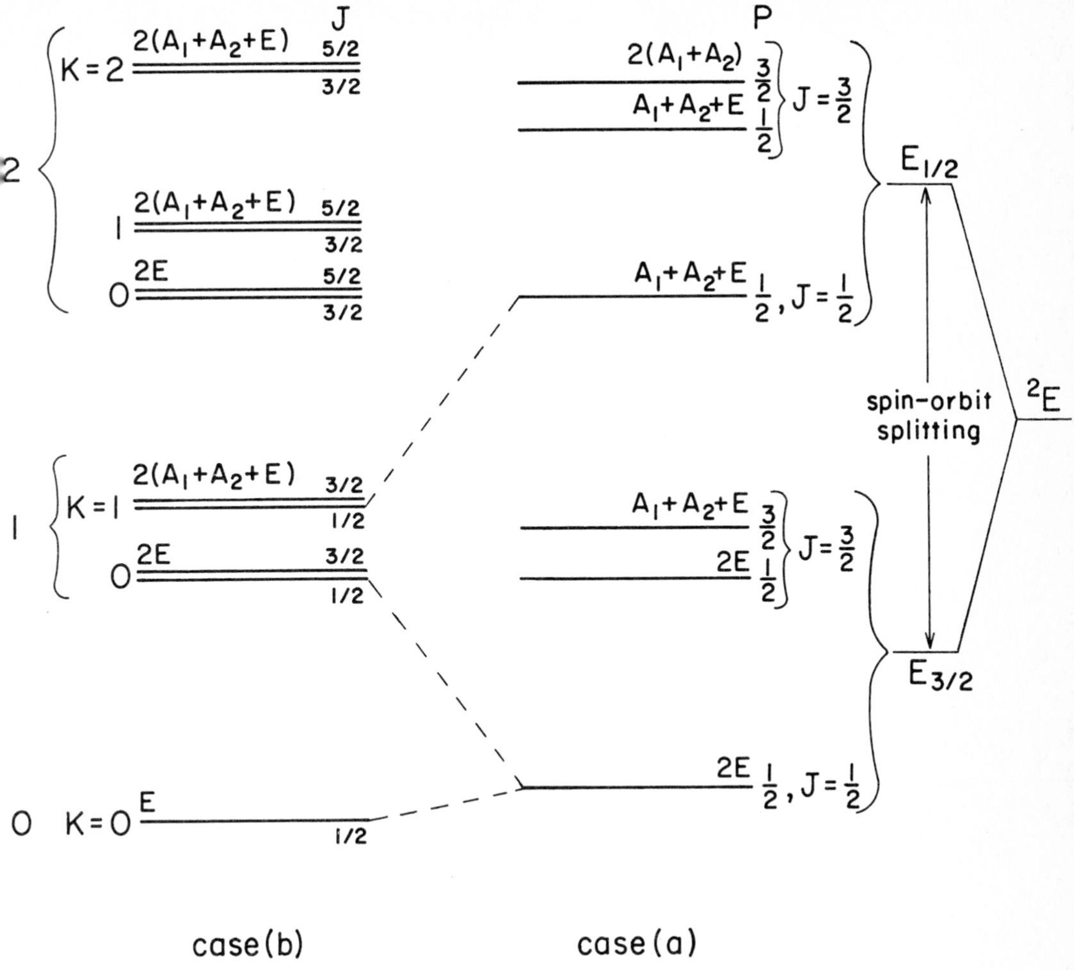

Fig. 3. The energy levels for CH_3F^+ in a ^2E vibronic state.

CH_3F^+

This ion has an odd number of electrons and it is necessary to consider two limiting cases of energy level pattern; the limiting cases are Hund's cases (a) and (b). In the case (b) limit it is considered that there is no magnetic coupling (spin-orbit coupling) between the odd electron and the nuclear framework and hence the electron spin wavefunction is invariant to any nuclear permutation. In the case (a) limit the electron spin is considered to be tightly coupled to the nuclear framework and to be transformed, therefore, by nuclear permutation.

Hund's case (b) limit. Using N as the rotational quantum number in case (b) (we have $\vec{J} = \vec{N} + \vec{S}$) the species of the rotational wavefunctions $S_{N,k,m}$ are the same as those of the $S_{J,k,m}$ of CH_3F. If we consider the situation when the molecule is in a ^2E electronic state and an A_1 vibrational state then the species of the rovibronic states in $C_{3v}(M)$ are as given down the left hand side of Fig. 3. These species are obtained from those in Fig.2 on multiplication by E and the extra double degeneracy comes from the electron spin doublet degeneracy. For doublet states $J = N \pm \frac{1}{2}$ and the J labels are as shown.

Hund's case (a) limit. If the coupling of the odd electron to the nuclear framework is strong (and this occurs when heavy atoms are present) then we must allow for the transformation of the electron spin wavefunctions by nuclear permutations. This presents us with the problem of dealing with the transformation properties of wavefunctions that involve half integral angular momentum. Rotational wavefunctions $S_{J,k,m}$ involving half integral angular momenta are changed in sign by a rotation of the molecule through 2π radians. Bethe showed that we can deal with this situation by introducing the fictitious operation R into $\underset{\sim}{K}$ (and, as we shall see, into the MS group) where **R** is the rotation of the molecule through 2π radians; R is considered as being distinct from the identity. A rotation through 4π radians is equivalent to the identity. The effect of introducing R is to double the size of $\underset{\sim}{K}$ (and the MS group), and we call

Table I. The $\underset{\sim}{C}_{3v}(M)$ Character Table

	E	(123) (132)	(12)* (23)* (31)*
A_1	1	1	1
A_2	1	1	-1
E	2	-1	0

Table II. The $\underset{\sim}{C}_{3v}(M)^2$ Character Table[a]

	E	(123)	(23)*	R	R(123)
	1	2	6	1	2
Eq.Rot[b]	R^0	$R_z^{2\pi/3}$	$R_{\pi/2}^{\pi}$	$R^{2\pi}$	$R_z^{8\pi/3}$
A_1	1	1	1	1	1
A_2	1	1	-1	1	1
E	2	-1	0	2	-1
$E_{\frac{1}{2}}$	2	1	0	-2	-1
$E_{\frac{3}{2}}$	2	-2	0	-2	2

sep

[a]Only one element from each class is shown but the number of elements in each class is given.

[b]See Eq. 4. for the definition of the Equivalent Rotations.

the double groups $\underset{\sim}{K}^2$ and, for example, $\underset{\sim}{C}_{3v}(M)^2$. We now look at the classification of the case (a) wavefunctions of CH_3F^+ in $\underset{\sim}{C}_{3v}(M)^2$; the character table of the $\underset{\sim}{C}_{3v}(M)^2$ group is given in Table II.

Using $\underset{\sim}{K}^2$ we classify the pair of $S = \frac{1}{2}$ electron spin functions as $D^{(\frac{1}{2})}$ and by correlation of the group $\underset{\sim}{K}^2$ with $\underset{\sim}{C}_{3v}(M)^2$ we determine that this pair of electron spin functions transform as $E_{\frac{1}{2}}$ in $\underset{\sim}{C}_{3v}(M)^2$. We consider the situation in which the electronic orbital function is of species E so that the electronic spin-orbital species is $E \times E_{\frac{1}{2}} = E_{\frac{1}{2}} + E_{\frac{3}{2}}$; strong spin-orbit coupling produces a large splitting between the $E_{\frac{1}{2}}$ and $E_{\frac{3}{2}}$ states. The rotational wavefunction $S_{J,p,m}$ (where it is conventional to use p for the z component of J in case (a)) can be classified in $\underset{\sim}{C}_{3v}(M)^2$; they transform as $E_{\frac{1}{2}}$ if $P = |p| = \left(\frac{3n}{2} \pm 1\right)$ and as $E_{\frac{3}{2}}$ if $P = \frac{3n}{2}$, where n is an odd integer. The final rovibronic-electron spin energy levels and symmetry labels in case (a) are obtained by multiplying the rotational and spin-orbit species, and the results are shown on the right hand side of Fig. 3. The correlation of the case (a) and case (b) limits can be carried out maintaining the $\underset{\sim}{C}_{3v}(M)$ and J symmetry labels; this is shown in Fig. 3 for the $J = \frac{1}{2}$ levels. Note that rovibronic-electron spin states always transform according to the simple group and not the double group of the MS group. Depending on the circumstances a given molecule would conform to some intermediate position on such a correlation diagram but would approach case (b) for high J values.

$CH_3BF_2^+$

For a non-rigid molecule with an odd number of electrons, such as $CH_3BF_2^+$, we do not have a single case (a) limit but two, depending upon which of the two internal rotors (see Fig. 4) the odd electron spin is coupled to. Furthermore there are two distinct electron spin double groups, one for each of these limiting cases.

We consider first case (a) in which the odd electron is

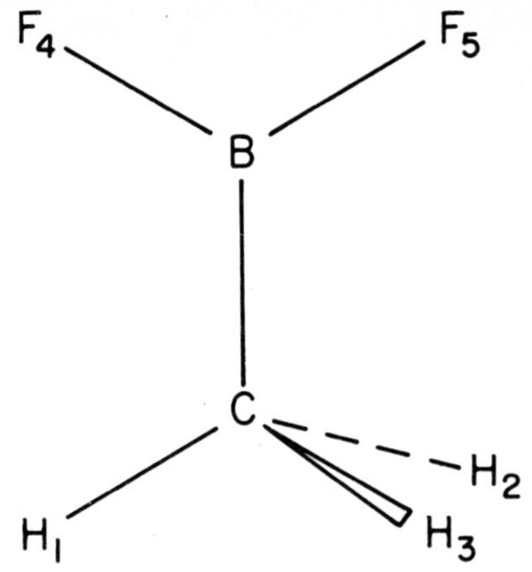

Fig. 4. The $CH_3BF_2^+$ molecule

Table III. The Character Table[a] of $\underset{\sim}{G}^2_{12}(CH_3)$ for $CH_3BF_2^+$

	E	(123)	(12)*	(45)	(123)(45)	(12)(45)*	R	R(123)	R(45)	R(123)(45)	
	1	2	6	1	2	6	1	2	1	2	
Eq.Rot	R^0	$R_z^{2\pi/3}$	$R^{\pi}_{\pi/6}$	R^0	$R_z^{2\pi/3}$	$R^{\pi}_{\pi/6}$	$R^{2\pi}$	$R_z^{8\pi/3}$	$R^{2\pi}$	$R_z^{8\pi/3}$	
A_1'	1	1	1	1	1	1	1	1	1	1	
A_2'	1	1	-1	1	1	-1	1	1	1	1	
E'	2	-1	0	2	-1	0	2	-1	2	-1	
A_1''	1	1	1	-1	-1	-1	1	1	-1	-1	
A_2''	1	1	-1	-1	-1	1	1	1	-1	-1	
E''	2	-1	0	-2	1	0	2	-1	-2	1	
$E_{\frac{1}{2}}'$	2	1	0	2	1	0	-2	-1	-2	-1	
$E_{\frac{1}{2}}''$	2	-1	0	-2	1	0	-2	1	2	-1	
$E_{\frac{3}{2}}'$	2	-2	0	2	-2	0	-2	2	-2	2	sep
$E_{\frac{3}{2}}''$	2	-2	0	-2	2	0	-2	2	2	-2	sep

[a]The z-axis is along C→B, the x-axis is in the H_1CB plane and the xyz axes are right handed.

Table IV. The Character Table* of $\underset{\sim}{G}^{2}_{12}(BF_2)$ for $CH_3BF_2^+$

	E	(123)	(12)*	(45)	(123)(45)	(12)(45)*	R	R(123)	R(123)(45)
	1	2	6	2	2	6	1	2	2
Eq.Rot	R^0	R^0	R^π_c	R^π_a	R^π_a	R^π_b	$R^{2\pi}$	$R^{2\pi}$	$R^{3\pi}_a$
A_1'	1	1	1	1	1	1	1	1	1
A_2'	1	1	-1	1	1	-1	1	1	1
E'	2	-1	0	2	-1	0	2	-1	-1
A_1''	1	1	1	-1	-1	-1	1	1	-1
A_2''	1	1	-1	-1	-1	1	1	1	-1
E''	2	-1	0	-2	1	0	2	-1	1
$E_{a/2}$	2	-1	0	0	$\sqrt{3}$	0	-2	+1	$-\sqrt{3}$
$E_{\frac{1}{2}}$	2	2	0	0	0	0	-2	-2	0
$E_{b/2}$	2	-1	0	0	$-\sqrt{3}$	0	-2	+1	$\sqrt{3}$

*The a-axis is along C-B, the b-axis is in the BF_2 plane perpendicular to the a-axis and the c-axis is perpendicular to both the a and b axes.

Table V. The Symmetry Species of the Torsional Wavefunctions ($e^{ik_i\rho}$) of $CH_3BF_2^+$.

K_i	Symmetry Species
0	A_1'
$6n\pm1$	E''
$6n\pm2$	E'
$6n\pm3$	$A_1''+A_2''$
$6n$	$A_1'+A_2'$

$K_i = |k_i|$ and n is integral.

tightly coupled to the "CH_3" end of the molecule (as might obtain in $SiCl_3BF_2^+$). In this situation the spin double group of the MS group is $\tilde{G}_{12}^2(CH_3)$ as given in Table III. We consider the situation for the molecule in a $^2E'$ electronic state. The $S = \frac{1}{2}$ electron spin functions transform as $D^{(\frac{1}{2})}$ in $\tilde{K}^2$ and hence as $E_{\frac{1}{2}}'$ in $\tilde{G}_{12}^2(CH_3)$. The electron spin-orbit states transform as $E_{\frac{1}{2}}' \times E' = E_{\frac{1}{2}}' + E_{\frac{3}{2}}'$. Using the equivalent rotations we find that the rotational functions $S_{J,p,m}$ transform as $E_{\frac{1}{2}}'$ if $P = \left(\frac{3n}{2} \pm 1\right)$ and as $E_{\frac{3}{2}}'$ if $P = \frac{3n}{2}$, where n is an odd integer. Multiplying the rotational and electron spin-orbit species together we determine the symmetry labels shown on the right hand side of Fig. 5. The case (b) levels (using CH_3 fixed xyz axes) and the (a)-(b) correlation are also shown in Fig. 5; this diagram is obviously very similar to Fig. 3 for CH_3F^+ and would be entirely appropriate for a molecule such as $SiCl_3BF_2^+$ where the $\tilde{C}_{3v}$ end is heavy.

For the case (a) situation in which the odd electron spin is coupled to the BF_2 group the spin double group is $\tilde{G}_{12}^2(BF_2)$, and the character table of this group is given in Table IV. Note the different equivalent rotations from those of $\tilde{G}_{12}^2(CH_3)$. The electron spin functions are of species $D^{(\frac{1}{2})}$ in $\tilde{K}^2$. from which we obtain $E_{\frac{1}{2}}$ in $\tilde{G}_{12}^2(BF_2)$, and thus the species of the electron spin-orbit states is given by $E_{\frac{1}{2}} \times E' = E_{a/2} + E_{b/2}$. We determine that the species of the pair of functions $S_{J,\pm p,m}$ is $E_{\frac{1}{2}}$ for p half integral so that the species of these case (a) levels are as given on the right hand side of Fig. 6. The case (b) levels (using BF_2 fixed axes) and the correlation are also shown in Fig. 6. This figure is appropriate for $CH_3BF_2^+$ since the BF_2 end is heavy.

The correlation of the levels in Fig. 5 with those in Fig. 6 require the introduction of the torsional quantum number k_i. The torsional wavefunction (assuming no torsional barrier) is given by $e^{ik_i\rho}$ where $k_i = 0, \pm 1, \pm 2, \ldots$ and ρ is the dihedral angle between the H_1-C-B and F_4-B-C planes. The symmetry species of the torsional wavefunctions are given in Table V. Using the results in Table V we can determine the species of the rotational-

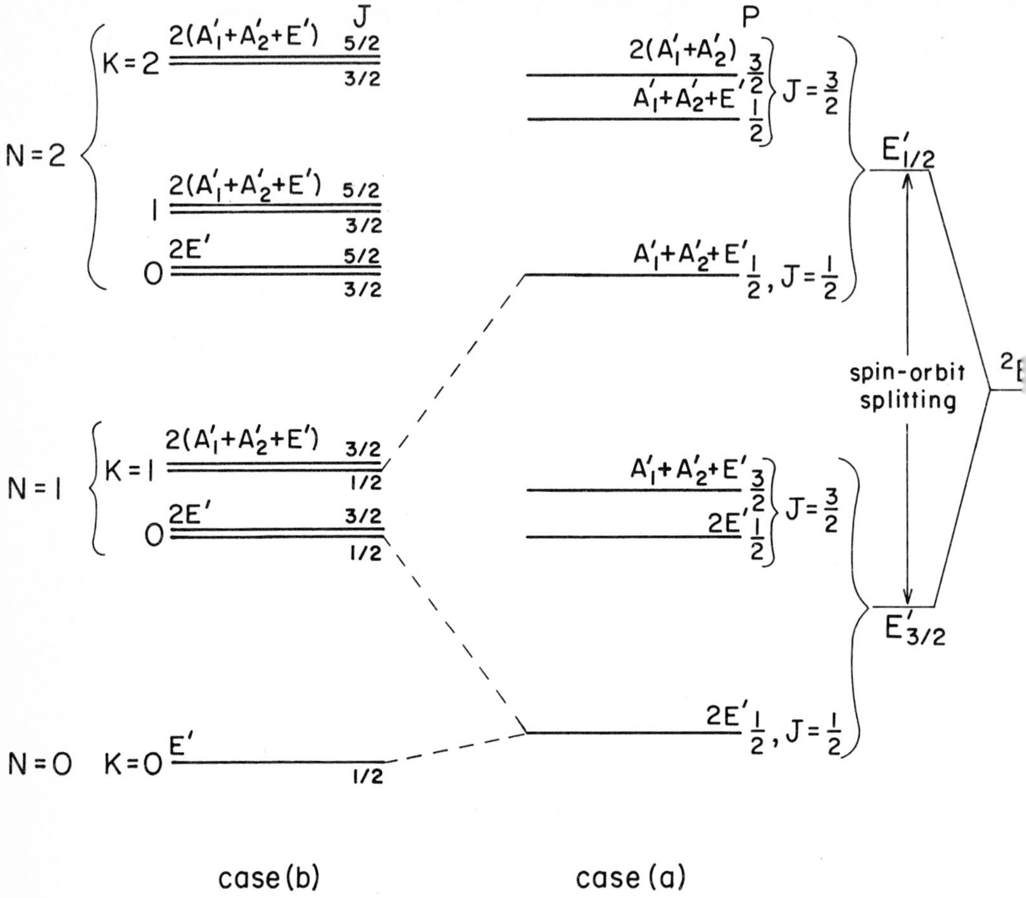

Fig. 5. The energy levels for $CH_3BF_2^+$ in a $^2E'$ vibronic state with CH_3 fixed axes.

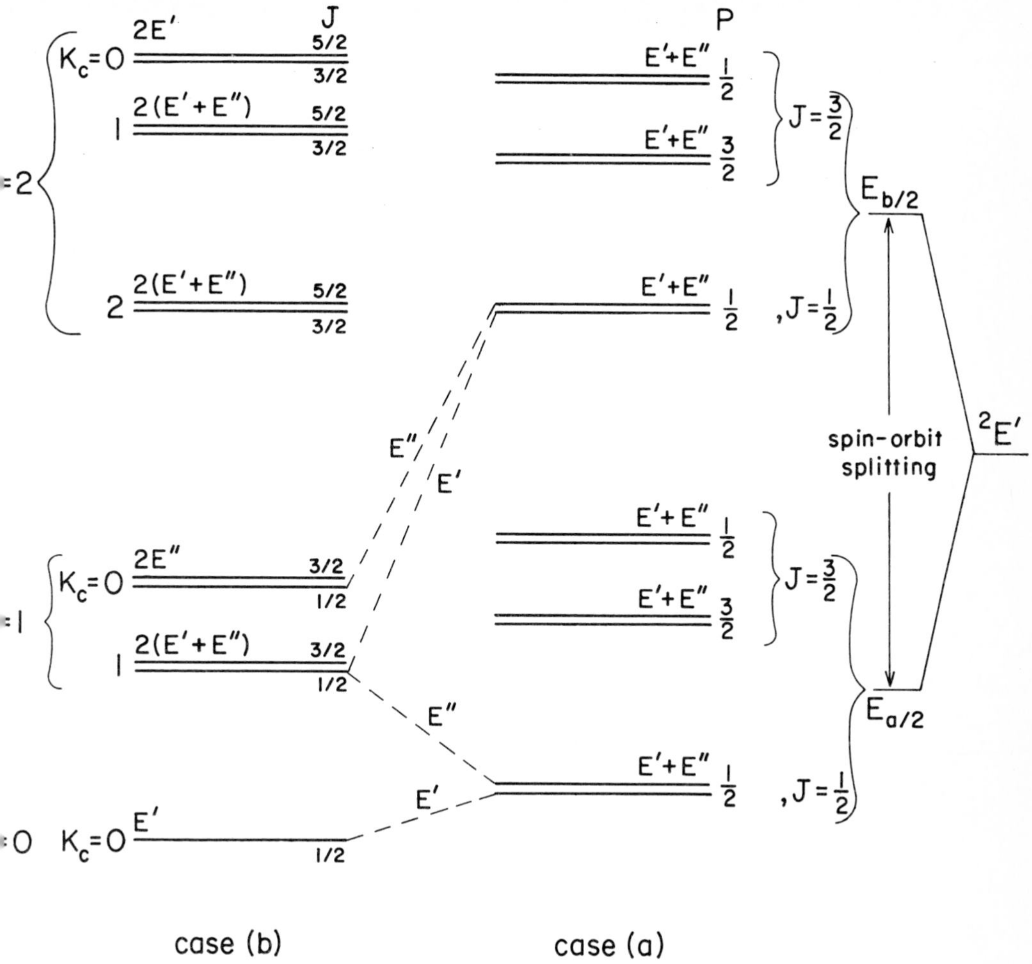

g. 6 The energy levels for $CH_3BF_2^+$ in a $^2E'$ vibronic state with BF_2 fixed axes.

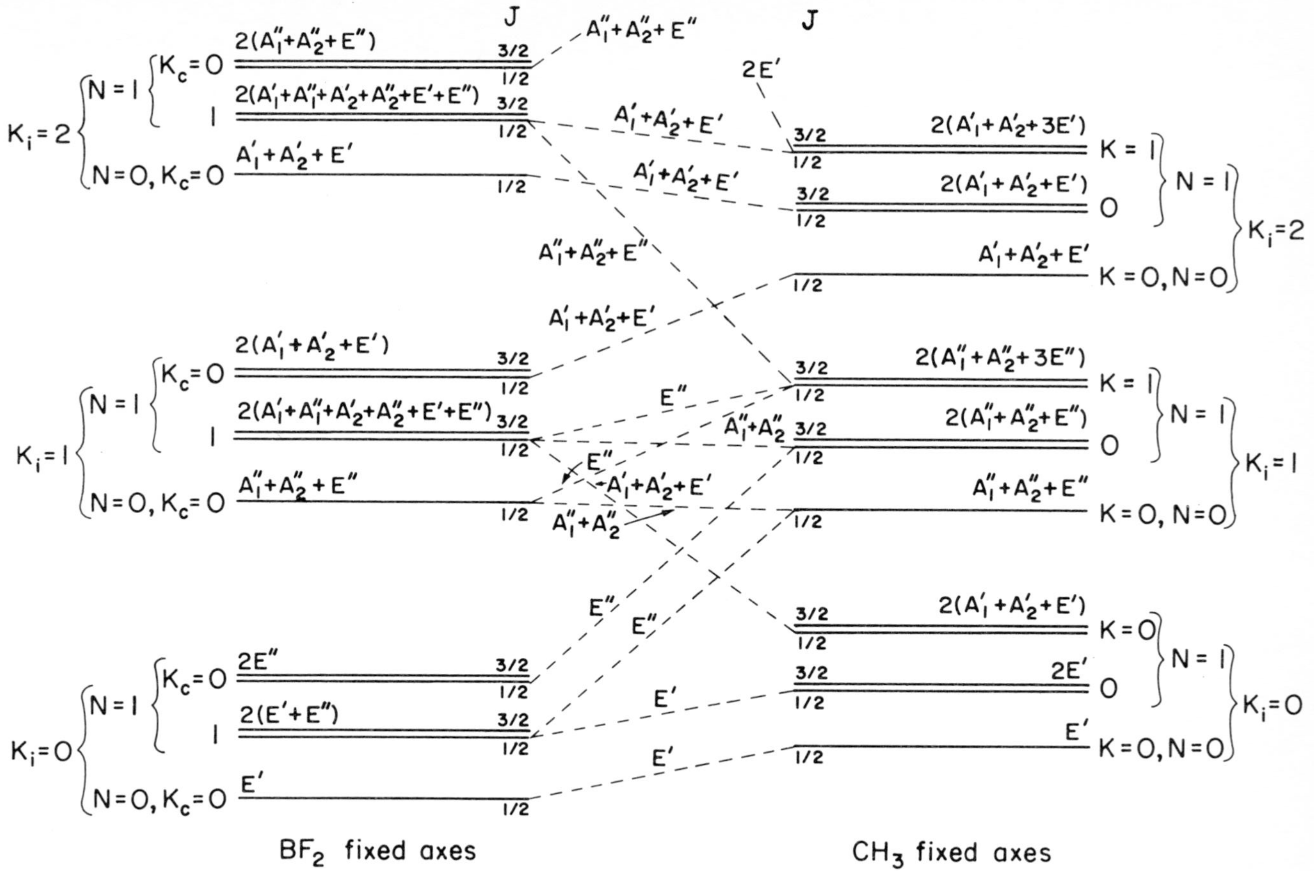

Fig. 7 The correlation of the case (b) limits for $CH_3BF_2^+$ where K_i is the torsional quantum number.

torsional-vibronic-electron spin states and the correlation of the two case (b) limits is given in Fig. 7. The case (a) levels can be similarly correlated. The correlation of these energy levels with those of a rigid $CH_3BF_2^+$ molecule, in which the torsional states are widely spaced vibrational states, can also be determined.

Concluding remarks

In this paper many different 'appropriate' approximate energy level patterns have been drawn and symmetry labelled using $\underset{\sim}{K}$ and the MS group. For a rigid molecule in an isolated singlet ground electronic state and having no resolved nuclear hyperfine structure the situation is rather straightforward as exemplified by the CH_3F case. For a rigid molecule with unpaired electron spin we are required to consider both the case (a) and case (b) limiting situations; CH_3F^+ provides an example of this. The most complicated situation we have considered, $CH_3BF_2^+$, is that of a non-rigid molecule in a non-singlet state. In this case there are many possible limiting energy level patterns depending on which rotor is the heavier, on the magnetic coupling of the odd electron spin, and on the height of the torsional barrier. Some of these limiting cases have been drawn, symmetry labelled and correlated.

We have only discussed here the problem of labelling the energy levels. These labels are of more than bureaucratic interest and enable us to draw the energy level correlations as well as to understand interactions between levels caused by neglected terms in the Hamiltonian and by externally applied perturbations. Of particular importance is the fact that electric dipole transitions in a molecule are only possible between states whose species are connected by $D^{(1)}$ (in $\underset{\sim}{K}$) and by Γ^* (in the MS group), where Γ^* is that one-dimensional irreducible representation of the MS group having character +1 for all nuclear permutations and character -1 for all nuclear permutations with the inversion. This results from the fact that the electric dipole moment operator is of species $D^{(1)}$ and Γ^*.

REFERENCES

The spin double group:

1. H. Bethe, Ann. der Physik (5) 3, 133 (1929).
2. W. Opechowski, Physica VII, 552 (1940).

The molecular symmetry group:

3. J.T. Hougen, J Chem. Phys., 37, 1433 (1962); 39, 358 (1963).
4. H.C. Longuet-Higgins, Mol. Phys., 6, 445 (1963).
5. J.K.G. Watson, Can. J. Phys., 43, 1996 (1965).
6. P.R. Bunker and D. Papousek, J. Mol. Spectrosc., 32, 419 (1969).
7. P.R. Bunker, Vibrational Spectra and Structure, 3, 1 (1975).
8. P.R. Bunker, 'Molecular Symmetry and Spectroscopy', Academic Press, in press.

RELATIONSHIP BETWEEN THE FEASIBLE GROUP AND THE POINT GROUP OF A RIGID MOLECULE

J. D. Louck†

Group T-7, Theoretical Division
Los Alamos Scientific Laboratory
Los Alamos, New Mexico 87545, USA

ABSTRACT

The rôle of the point group in conventional rigid molecule theory is reviewed and its relationship to the feasible group discussed.

It is very important to understand the relationship between the feasible group of a rigid molecule as discussed in Dr. Hougen's talk and the conventional point group of the molecule. It is my belief that neither of these groups is more fundamental than the other, that both concepts generalize to nonrigid molecules, and that both groups (as well as others) are important in the study of molecular models. This viewpoint will be defended here only for rigid molecules, and in such a way as to complement Dr. Hougen's presentation.

The viewpoint I shall present is the conventional one and is developed, for example, in the article by Wilson and Howard (1) and in the book by Wilson, Decius, and Cross (2). I shall, however, emphasize throughout this talk the rôles of the point group and of the moving frame. My colleague, Harold W. Galbraith, and I have developed these details elsewhere (Ref. 3,4), but I believe it to be useful to review again the more essential features of this approach. I shall focus on two aspects of the description of a rigid molecule: (i) the description of the static model (equilibrium configuration) of a molecule; and (ii) the description of the motion of the dynamical model of the molecule in space-time.

Consider first the description of the static model (the dumbbell model made up of rods and spheres). The static model will be described in a laboratory frame L with basis vectors $(\hat{\ell}_1, \hat{\ell}_2, \hat{\ell}_3)$ (a right-handed triad of perpendicular unit vectors) which is a principal axes system located at the center of mass. Let A denote the set of vectors

$$A = \{ \vec{a}^{\alpha} | \alpha = 1,2,\ldots N \} , \tag{1}$$

where $\vec{a}^{\alpha}$ is the position vector from the origin of L to the point where the atom labelled by α is located. Each vector $\vec{a}^{\alpha}$ may be expressed relative to the frame L as

$$\vec{a}^{\alpha} = a_1^{\alpha} \hat{\ell}_1 + a_2^{\alpha} \hat{\ell} + a_3^{\alpha} \hat{\ell}_3 , \tag{2}$$

where $(a_1^{\alpha}, a_2^{\alpha}, a_3^{\alpha})$ are specified real numbers.

Consider next the partitioning of A corresponding to sets of identical particles. Let A_k denote the subset of A consisting of position vectors of identical particles of "type k". Then A may be written as the union of the disjoint subsets $\{ A_k | k \varepsilon K \}$, where K is a set indexing the distinct types of atoms:

$$A = \bigcup_k A_k \quad . \tag{3}$$

We now define the point group G of a molecule with static model A:

$$G = \{ \mathbf{R} | R \varepsilon G \text{ and } R: A_k \rightarrow A_k, \text{ each } k \varepsilon K \} , \tag{4}$$

where G denotes the group of rotation-inversions of the space R^3 [Euclidean 3-space with points (x_1, x_2, x_3) which we will describe using the frame L].

We shall use the notation g to denote an element of the point group G.

There are two representations of the group G which play a significant rôle in this presentation:

(i) The representation of a proper rotation g as a linear transformation of the points of R^3. In vector notation, we have

$$g: \ R^3 \rightarrow R^3$$

$$\vec{x} \rightarrow \vec{y} = g \vec{x} = \vec{x} \cos \phi + (\hat{n} \cdot \vec{x}) \vec{x} (1 - \cos \phi) + (\hat{n} \times \vec{x}) \sin \phi , \tag{5}$$

where g is a positive rotation (right-hand screw rule) by angle ϕ about the direction specified by the unit vector $\hat{n}$ as determined from $g = R \varepsilon G$. g may also be represented by the 3x3 proper orthogonal matrix with element in row i and column j given by

$$R_{ij}(t) = \hat{\ell}_i \cdot g \hat{\ell}_j \quad . \tag{6}$$

The inversion I of the space R^3 is defined by $I \vec{x} = - \vec{x}$ and is represented by $- I_3$, where I_3 denotes the 3x3 unit matrix.

(ii) The representation of g as a linear transformation on the elements of the set A. We may write

$$g: A \to A \tag{7}$$

$$[\vec{a}^1\vec{a}^2 \ldots \vec{a}^N] \to [\vec{a}^1\vec{a}^2 \ldots \vec{a}^N]\, P(g)$$

where we have ordered the elements of A and placed them in a 1xN row vector. P(g) is then an NxN permutation matrix.

Observe that the group multiplication properties are satisfied:

$$g'\ (g\vec{x}) = (g'g)\vec{x}\ .\ \text{each}\ \vec{x} \in R^3\ ,$$

$$R(g')R(g) = R(g'g)\ , \tag{8}$$

$$P(g')P(g) = P(g'g)\ ,$$

for all g,g' $\in$ G. Thus, the two correspondences

$$g \to R(g) \text{ and } G \to P(g),\ \text{each } g \in G, \tag{9}$$

are representations of the point group G.

If we denote by A the 3xN matrix

$$A = \begin{bmatrix} a_1^1\ a_1^2\ \ldots\ a_1^N \\ a_2^1\ a_2^2\ \ldots\ a_2^N \\ a_3^1\ a_3^2\ \ldots\ a_3^N \end{bmatrix}, \tag{10}$$

then A intertwines the representations { P(g) P $\in$ G} , that is,

$$R(g)\ A = A\ P(g),\ \text{each}\ \ g \in G. \tag{11}$$

Relation (11) is the key result obtained from the static model of a rigid molecule.

Consider next the model for the motion of the molecule in space-time. Intuitively, we have in mind the following situation. We imagine that the rigid framework translates and rotates in space and that the atoms execute small oscillatory motions in the neighborhood of the (moving) equilibrium points. This intuitive picture for a set of motions of N particles corresponds to our conception of the motions (based on empicical knowledge) of what is today called a "rigid" molecule. There are sufficiently many molecules of the "rigid type" to justify a careful development of such a model. [A phenomenological model of a molecule such as this one clearly ignores many aspects of a "real molecule," and one does not expect the model to have general validity - the model is designed specifically for the description of vibtation-rotation motions of the atoms, and even then for a limited energy domain.]

Even after settling on the model above, there are still many approaches that one might use to formulate a description of the motions. Let us continue the intuitive discussion. The use of a moving reference frame is suggested if one wishes to obtain a Hamiltonian for the system which, for motions in the neighborhood of the equilibrium configuration, has the approximate form

$$T_{C.M.} + H_R + H_V , \tag{12}$$

where $T_{C.M.}$ is the kinetic energy of the center of mass, H_R the rotational energy, and H_V a Hamiltonian term for the kinetic and potential energies of the small motions near equilibrium. Intuition suggests the following forms for H_R and H_V:

$$\begin{aligned} H_R &= \frac{J_1^2}{2I_1} + \frac{J_2^2}{2I_2} + \frac{J_3^2}{2I_3} , \\ H_V &= \sum_{\mu=1}^{3N-6} (p_\mu^2 + \omega_\mu^2 q_\mu^2)/2. \end{aligned} \tag{13}$$

In the definition of H_R, the symbol J_i (i=1,2,3) denotes the component of the total angular momentum $\vec{J}$ along the i-th axis of the moving frame; I_i is the principal moment of inertia of the equilibrium configuration about the i-the axis of the moving frame. (We choose the moving frame to coincide with a principal axis system when the atoms are located at their equilibrium points.) In the definition of H_V, the symbol q_μ (μ=1,2,..., 3N-6) denotes a normal coordinate, p_μ its conjugate linear momentum, and ω_μ the frequency of the μ-th normal mode of oscillation. (The normal mode analysis of the vibrational motion may be carried out on the non-rotating molecule by several available methods.)

Let us next consider how one may give a precise formulation of the approach outlined above.

The first problem which must be solved is that of finding an appropriate moving reference frame. Eckart (5) solved this problem by imposing two conditions on the moving frame:

(i) Casimir's condition. In the limit of vanishing displacements away from the equilibrium configuration, the Coriolis interaction between rotation and internal motions should be zero.

(ii) Linearity of internal coordinates. The internal degrees of freedom should be described by coordinates which are linear combinations (with fixed numerical coefficients) of the components of the displacements away from equilibrium where the components are referred to the moving frame.

The second condition is imposed to assure that the normal coordinates calculated for the non-rotating molecule can be carried over, without change, to the rotating, vibrating molecule.

[We have gone to the trouble of briefly discussing the elementary (and standard) results above because the Eckart frame is the key concept for understanding the rôle of the point group G in the molecular motions problems for rigid molecules.]

The explicit construction of the Eckart frame may be given in the following manner: (i) Introduce the three vectors $\vec{F}_i$ defined by

$$\vec{F}_i \equiv \sum_\alpha m_\alpha a_i^\alpha \vec{x}^\alpha , \tag{14}$$

where $\vec{x}^\alpha$ is the instantaneous position vector of atom α in the laboratory frame L. (ii) Define a triad of unit perpendicular vectors $(\hat{f}_1, \hat{f}_2, \hat{f}_3)$ by

$$\hat{f}_i = \sum_j \left(F^{-1/2}\right)_{ij} \vec{F}_j , \tag{15}$$

where F denotes the symmetric Gram matrix with elements $F_{ij} = \vec{F}_i \cdot \vec{F}_j$. [We assume that $\vec{F}_1, \vec{F}_2, \vec{F}_3$ are linearly independent so that F is positive definite; $F^{-1/2}$ is then, by definition, the positive definite matrix such that $F^{-1/2} F^{-1/2} = F^{-1}$.] The three vectors $(\hat{f}_1, \hat{f}_2, \hat{f}_3)$ then define a moving frame F such that conditions (i) and (ii) above are satisfied.

Observe that the Eckart vectors depend implicitly on the particle position vectors $\vec{x}^{\alpha}$. If we denote this result by writing $\hat{f}_i(\vec{x}^1, \ldots, \vec{x}^N)$, then one easily verifies from Eqs. (14) and (15) that

$$\hat{f}_i(R\vec{x}^1, \ldots, R\vec{x}^N) = (R\hat{f}_i)(\vec{x}^1, \ldots, \vec{x}^N), \text{ each } R \in G, \tag{16}$$

$$\hat{f}_i(\vec{x}^1 + \vec{a}, \ldots, \vec{x}^N + \vec{a}) = \hat{f}_i(\vec{x}^1, \ldots, \vec{x}^N) \text{ for arbitrary translations } \vec{a}. \tag{17}$$

Thus, under an arbitrary rotation-inversion R of the molecule, the Eckart frame undergoes the same rotation-inversion. Under an arbitrary translation of the molecule, the Eckart frame is invariant. (It is this second property which allows us to consider that the Eckart frame is located at the center of mass of the moving molecule.)

Now that the moving reference frame F is defined (we consider only nonplanar molecules here), we may determine the transformation from the Cartesian coordinates $x_i^{\alpha} = \vec{x}^{\alpha} \cdot \ell_i$ relative to the laboratory frame to "molecular coordinates." The position vector of atom α is given by

$$\vec{x}^{\alpha} = \vec{R} + \vec{c}^{\alpha} + \vec{\rho}^{\alpha} , \tag{18}$$

where

(i) $\vec{R}$ is the instantaneous center of mass vector;

(ii) $\vec{c}^{\alpha}$ is the position vector of the moving equilibrium point of atom α relative to the center of mass and of the form

$$\vec{c}^{\alpha} = a_1^{\alpha}\hat{f}_1 + a_2^{\alpha}\hat{f}_3 + a_3^{\alpha}\hat{f}_3 \tag{19}$$

when referred to the moving Eckart frame;

(iii) $\vec{\rho}^{\alpha}$ is the displacement vector of atom α from the equilibrium point $\vec{R} + \vec{c}^{\alpha}$.

There are six conditions imposed on the displacement vectors $\{\vec{\rho}^{\alpha}\}$:

Center of mass condition: $$\sum_{\alpha} m_{\alpha}\vec{\rho}^{\alpha} = \vec{0}, \tag{20}$$

Eckart conditions: $$\sum_{\alpha} m_{\alpha}\, \vec{c}^{\alpha} \times \vec{\rho}^{\alpha} = \vec{0}. \tag{21}$$

[The second set of conditions results from an easily proved relation between the Eckart frame vectors $\hat{f}_i$ and the $\vec{F}_i$:

$$\hat{f}_1 \times F_1 + \hat{f}_2 \times F_2 + \hat{f}_3 \times \vec{F}_3 = \vec{0} .]$$

Let us call a set of vectors $\{\vec{\rho}^{\alpha} \mid \alpha = 1,\ldots,N$ which satisfies Eqs. (20) and (21) a set displacement vectors for the frame $(\hat{f}_1,\hat{f}_2,\hat{f}_3)$.

It is convenient to reformulate Eqs. (20) and (21) in the form:

$$\sum_{\alpha} \vec{s}_i^{\alpha}\cdot\vec{\rho}^{\alpha} = 0,\ (i = 1,2,3),\ (20')$$

$$\underset{\alpha}{} \vec{s}_i^{\alpha}\cdot\vec{\rho}^{\alpha} = 0,\ (ib= 4,5,6),\ (21')$$

where

$$\vec{s}_i^{\alpha} = m_{\alpha}\hat{f}_i/m^{1/2} \quad , m = \sum_{\alpha} m_{\alpha} \quad ,$$

$$\vec{s}_{i+3}^{\alpha} = m_{\alpha}N_i^{-1/2}(\hat{f}_i \times \vec{c}^{\alpha}) \tag{22}$$

$$N_i = \sum_{\alpha} m_{\alpha}\, (\hat{f}_i \times \vec{c}^{\alpha})\cdot(\hat{f}_i \times \vec{c}^{\alpha})$$

for $i = 1,2,3$. Observe that the vectors in the set $\{\vec{s}_t^{\alpha} \mid t = 1,\ldots,6\}$ satisfy the orthogonality relations:

$$\sum_{\alpha} m_{\alpha}^{-1}\vec{s}_r^{\alpha}\cdot\vec{s}_t^{\alpha} = \delta_{rt} \quad . \tag{23}$$

It is always possible (in infinitely many ways) to find additional vectors $\vec{s}^{\alpha}_{\mu+6}$, $\mu = 1,2,\ldots,3N-6$, such that the 3N vectors in the set

$$\{ \vec{s}^{\alpha}_{t} | t = 1,\ldots,3N\} \tag{24}$$

satisfy the orthogonality relations

$$\sum_{\alpha} m_{\alpha}^{-1} \vec{s}^{\alpha}_{r} \cdot \vec{s}^{\alpha}_{t} = \delta_{rt}. \tag{25}$$

Furthermore, all vectors $\vec{s}^{\alpha}_{t}$ may be chosen to have numerical components relative to the Eckart frame, that is,

$$s^{\alpha}_{t,i} \equiv \vec{s}^{\alpha}_{t} \cdot \hat{f}_{i} = \text{real numerical constant} . \tag{26}$$

These results all follow from the fact that it is possible to construct a 3Nx3N orthogonal matrix with numerical entries in infinitely many ways when only the first six rows are specified (numerical) row vectors ($N > 3$).

The principal result obtained from the above analysis is: The rotation-inversion invariants defined by

$$Q_{\mu} = \sum_{\alpha} \vec{s}^{\alpha}_{\mu+6} \cdot \vec{\rho}^{\alpha} , \quad \mu = 1,2,\ldots,3N-6 \tag{27}$$

span the 3N-6 dimensional space of the internal coordinates. [We use the term "span" here in the same sense that each internal coordinate of the form $\xi = \sum_{\alpha} \vec{\xi}^{\alpha} \cdot \vec{\rho}^{\alpha}$, $\vec{\xi}^{\alpha}$ a triple of real numbers, has the form $\xi = \sum_{\mu} \xi_{\mu} Q_{\mu}$. Furthermore, $\sum_{\mu} a_{\mu} Q_{\mu} = \sum_{\mu} b_{\mu} Q_{\mu}$ implies $a_{\mu} = b_{\mu}$.]

Using Eqs. (20'), (21') and (25), we may invert Eq. (27) to obtain

$$\vec{\rho}^{\alpha} = m_{\alpha}^{-1} \sum_{\mu} \vec{s}^{\alpha}_{\mu+6} Q_{\mu} . \tag{28}$$

Taking components of Eq. (18) relative to the laboratory frame, we obtain the transformation:

$$x_i^\alpha = R_i + \sum_j C_{ij}(a_j^\alpha + \rho_j^\alpha) \quad ,$$

$$\rho_j^\alpha = m_\alpha^{-1} \sum_\mu s_{\mu+6,i}^\alpha Q_\mu \quad . \tag{29}$$

For each choice of the vectors $\vec{s}_{\mu+6}^{\alpha}(\mu = 1,\ldots,3N-6)$, Eqs. (29) define an explicit transformation from the 3N Cartesian coordinates $\{ x_i^\alpha \}$ to the 3N coordinates

$$R_i(i = 1,2,3);\ C \text{ (containing 3 independent coordinates)};$$
$$Q = \{ Q_\mu | \mu = 1,\ldots,3N-6 \} . \tag{30}$$

Furthermore, the transformation (29) is invertible for those values of the $\{ x_1^\alpha\}$ for which the Eckart frame construction exists (det $F \neq 0$).

This completes the construction of the dynamical molecular model. Using the transformation (29) the classical Hamiltonian

$$H = \frac{1}{2} \sum_{\alpha,i} m_\alpha (\dot{x}_i^\alpha)^2 + V(x_i^\alpha) \tag{31}$$

and the quantum mechanical Hamiltonian

$$H_{\sigma p} = -\frac{1}{2} \sum_\alpha \left(\frac{\partial}{\partial x_i^\alpha}\right)^2 + V(x_i^\alpha) \tag{32}$$

may be transformed unambiguously to the 3N molecular coordinates (30) (confer Ref. 4). Furthermore, when the potential energy is approximated by a quadratic form in the internal coordinates $\{ Q_\mu \}$, one obtains from Eqs. (31) and (32) approximate Hamiltonians of the forms (13).

We are now in a position to state the rôle of the point group G in the dynamical molecular model. We first define the action of the group G on a generic set $Y = \{ \vec{y}^\alpha | \alpha = 1,2,\ldots,N \}$ of instantaneous position vectors relative to the center of mass: $\vec{y}^\alpha \equiv \vec{x}^\alpha - \vec{R}$. Consider the set of linear operators

$$L(G) = \{ L_g | g \in G \} , \tag{33}$$

where $L_g: Y \to Z$ is the linear mapping of a set Y of instantaneous position vectors (relative to the center of mass) onto a second set Z of instantaneous position vectors (relative to the center of mass) given by

$$L_g: \vec{y}^{\alpha} \to \vec{z}^{\alpha} = L_g\vec{y}^{\alpha} = \sum_{\beta} (g\vec{y}^{\beta})P_{\alpha\beta}(g) \tag{34}$$

in which

(i) $\{P(g)\ |g \in G\}$ is the NxN permutation matrix representation (7) of G;

(ii) $g\vec{y}^{\beta}$ is the linear transformation defined for an arbitrary vector $\vec{y}$ by

$$g\vec{y} = \sum_{ij} R_{ij}(g)\ (\vec{y}\cdot\hat{f}_j)\hat{f}_i\ , \tag{35}$$

where $\{R(g)|\ g \in G\}$ is the 3x3 orthogonal matrix representation (6) of G, and the Eckart frame vectors $\hat{f}_i$ $(i = 1,2,3)$ are those corresponding to position vectors $\vec{y}^1,\ldots,\vec{y}^N$, that is, $\hat{f}_i = \hat{f}_i(\vec{y}^1 \ldots,\vec{y}^N)$. Observe that $g\vec{y}^{\alpha}\cdot g\vec{y}^{\beta} = \vec{y}^{\alpha}\cdot\vec{y}^{\beta}$ so that g is a rotation-inversion and that $g'\ (g\vec{y}) = (g'g)\vec{y}$.

An alternative expression for the transformation (34) in terms of the components $\vec{y}^{\alpha}\cdot\hat{f}_i$ and $(L_g\vec{y}^{\alpha})\cdot\hat{f}_i$ relative to the Eckart frame vectors $\hat{f}_i = \hat{f}_i\ (\vec{y}^1,\ldots,\vec{y}^N)$ is

$$(L_g\vec{y}^{\alpha})\cdot\hat{f}_i = \sum_{\beta,j} [P(g) \otimes R(g)]_{\alpha i;\beta j}\ \vec{y}\cdot\hat{f}_j\ , \tag{36}$$

where $P(g)$ $R(g)$ is the (matrix) direct product of $P(g)$ with $R(g)$ and $[P(g)\otimes R(g)]_{\alpha i\cdot\ \beta j} = P_{\alpha\beta}(g)\ R_{ij}(g)$ denotes the element of $P(g)$ $R(g)$ in row αi and column βj.

The operators $\{L_g|g \in G\}$ satisfy the following relations:

(i) $L_{g'}(L_g\vec{y}^\alpha) = L_{g'g}\vec{y}^\alpha$ for arbitrary $\vec{y}^\alpha$;

(ii) $L_g\vec{c}^\alpha = \vec{c}^\alpha$, $\alpha = 1,2,\ldots,N$; (37)

(iii) $L_g\hat{f}_i = \hat{f}_i$, $i = 1,2,3$,

where

$$(L_g\hat{f}_i)(\vec{y}^1,\ldots,\vec{y}^N) \equiv \hat{f}_i(L_g\vec{y}^1,\ldots,L_g\vec{y}^N).$$

The proofs of these relations are straightforward and may be found in Ref. 3.

Equation (i) means that the correspondence $g \to L_g$ is a linear representation of G; Eq. (ii) means that G is an isotropy group of the set of vectors $\{\vec{c}^\alpha|\alpha = 1,\ldots,N\}$ which define the static model [Eq. (11) for the static model provides the proof of (ii)]; and Eq. (iii) means that the Eckart frame is invariant under the action L_g of G.

Equations (37) are the key relations for establishing the rôle of the point group in the dynamical molecular model.

A principal result is: The group of operators

$$L(G) = \{L_g|g \in G\} \tag{38}$$

may be used to split the space of internal coordinates into subspaces which transform irreducibly under L(G).

Proof. Let $\{\rho^\alpha\}$ denote any set of displacement for the frame $(\hat{f}_1,\hat{f}_2,\hat{f}_3)$. Since $L_g\vec{y}^\alpha = \vec{c}^\alpha + L_g\vec{\rho}^\alpha$, it follows that the set of vectors

$$\{L_g\vec{\rho}^\alpha|\alpha = 1,2,\ldots,N\} \tag{39}$$

is also a set of displacement vectors for the frame $(\hat{f}_i,\hat{f}_2,\hat{f}_3)$.

Now choose any basis set $\{Q_\mu|\mu = 1,\ldots,3N-6\}$ for the internal coordinates [cf. Eqs. (24)-(28)]. Since

$$(L_g\vec{\rho}^{\alpha})\cdot\hat{f}_i = \sum_{\beta j} [P(g) \otimes R(g)]_{\alpha i;\beta j}\rho^{\beta}_j \quad , \tag{40}$$

we find

$$L_g Q_\mu = \sum_{\alpha} \vec{s}^{\alpha}_{\mu+6} \cdot (L_g\vec{\rho}^{\alpha}) = \sum_{\nu} M_{\mu\nu}(g)Q_\nu \quad , \tag{41}$$

where

$$M_{\mu\nu}(g) = \sum_{\alpha i\beta j} m_\beta^{-1}\, s^{\alpha}_{\mu+6,i}\, s^{\beta}_{\nu+6,j}\, [P(g) \otimes R(g)]_{\alpha i;\beta j} \quad . \tag{42}$$

The properties (i) $L_{g'g}$ (group property) and (ii) $\sum_\mu a_\mu Q_\mu = \sum_\mu b_\mu Q_\mu$ implies $a_\mu = b_\mu$ (basis property) together imply that the set of matrices (dimension 3N-6)

$$\{M(g) | g \in G\} \tag{43}$$

is a representation of G. The complete reduction of this representation into irreducible representations of G then defines internal coordinates of the form

$$\xi^{\Gamma}_{\gamma} = \sum_{\mu} a^{\Gamma}_{\gamma\mu}\, Q_\mu , \ \gamma = 1,2,\ldots,\dim \Gamma , \tag{44}$$

which are transformed according to irreducible representation Γ of G under the action L_g of G.

We see from the above that the group of operators L(G) solves fully the proplem of classifying the internal coordinates according to their transformation properties under the irreducible representations of the point group G.

Let us next consider the rôle of the permutation group in the molecular model. The Schrödinger equation for a molecule (point atoms + point electrons + Coulomb interactions) is invariant under any permutation of the coordinates of identical particles. The model of a molecule which we have described will, in general, not be invariant under all permutations of identical atoms; this is because the general permutational symmetry may be broken by the choice of a phenomenological potential energy function which, in principle, allows one to distinguish spatially, in a finite time, between atoms that would otherwise be called identical. Said somewhat differently, the form of the phenomenological potential energy function might not admit tunneling effects, hence, no tunneling will be predicted by such a model. Before any molecular model is complete, one must choose (by specifying the potential energy function) which permutations of coordinates of identical atoms are to be allowed and which are not. (A bad choice may give a model whose predictions do not agree well with experiment.)

In the model of a "rigid" molecule, one implicitly assumes that no tunneling takes place. (This assumption is made when one chooses a potential energy function which keeps the atoms near their equilibrium positions.) The question then arises as to which permutations of coordinates of identical atoms are allowed by this model. The answer is easily given. Since the kinetic energy is invariant under the group P of all permutations of coordinates of identical atoms, it is the properties of the potential energy function alone under permutations of coordinates of identical atoms which restricts the symmetry of the Hamiltonian to a subgroup of P. But, by assumption, the potential energy function V of a rigid molecule is the most general function of the internal coordinates $\{Q_\mu = 1,\ldots,3N-6\}$ which is analytic in the neighborhood of the equilibrium configuration and such that:

(i) Equilibrium conditions

$$\left.\frac{\partial V}{\partial Q_\mu}\right|_{\text{equilibrium}} = 0; \tag{45}$$

(ii) Invariance conditions

$$V(L_g Q_1,\ldots,L_g Q_{3N-6}) = V(Q_1,\ldots,Q_{3N-6}), \tag{46}$$

each $g \in G$. Thus, our problem is to determine the properties of the function (46) under permutations of coordinates of identical atoms.

In order to examine the above problem, we require a careful definition of "permutations of coordinates of identical particles" and the resulting properties of the internal coordinates under such permutations.

Let there be N_k identical atoms of type k labelled by distinct integers

$$\alpha_1(k),\ \alpha_2(k),\ldots,\alpha_{n_k}(k)\ ,$$

and let the position vectors of these atoms in the laboratory frame be the vectors in the set X_k given by

$$X_k = \{\vec{x}^\alpha | \alpha = \alpha_1(k),\ldots,\alpha_{n_k}(k)\}\ , \tag{47}$$

where $\vec{x}^\alpha$ is the position vector of the atom labelled α. A permutation P_k of the set of position vectors of the identical atoms of type k is a mapping of X_k onto X_k. The product of two permutations P_k' and P_k is then the usual composition of mappings. The set of all such mappings then forms a group isomorphic to the symmetric group S_{n_k}. Indexing the "types" of identical atoms by $k = 1,2,\ldots,m$ ($\sum_{k=1}^{m} = N$), we see that $P_k P_{k'}$ $= P_k P_{k'} = P_{k'} P_k$ $(k' = k)$ on the set $X = \cup X_k$ and that each permutation $P\colon X \to X$ of the position vectors of identical atoms of the molecule has the form

$$P = \prod_{k=1}^{m} P_k\ , \quad P_k \in S_{n_k}\ . \tag{48}$$

Consider next the linear mapping L_g defined by Eq. (34). Then the linear mapping $P_g\colon Y \to Y$ defined by

$$P_g = g\, L_{g^{-1}}\ , \text{ each } g \in G \tag{49}$$

is a permutation of position vectors (relative to the center of mass) of identical atoms:

$$P_g\colon \vec{y}^\alpha \to P_g\vec{y}^\alpha = \sum_\beta \vec{y}^\beta\ P_{\beta\alpha}\ (g)\ . \tag{50}$$

Furthermore, the set of permutations which map Y onto Y given by

$$P(G) = \{\ P_g | g \in G\ \} \tag{51}$$

forms a group under composition of mappings; hence,

$$P_{g'}(P_g\vec{y}^\alpha) = P_{g'g}\vec{y}^\alpha\ . \tag{52}$$

The action of the group of permutations $P(G)$ on the vectors $\{\vec{c}^\alpha|\alpha = 1,\ldots,N\}$ which give the equilibrium positions relative to the Eckart frame and the set of displacement vectors $\{\vec{\rho}^\alpha|\alpha = 1,\ldots,N\}$ is given respectively by

$$P_g\vec{c}^\alpha = g\vec{c}^\alpha = \sum_\beta \vec{c}^\beta P_{\beta\alpha}(g) \quad , \tag{53}$$

$$P_g\vec{\rho}^\alpha = gL_{g^{-1}}\vec{\rho}^\alpha = \sum_\beta \vec{\rho}^\beta P_{\beta\alpha}(g) \quad . \tag{54}$$

Observe then that Eq. (52) is also valid when $\vec{y}^\alpha$ is replaced by $\vec{c}^\alpha$ or $\vec{\rho}^\alpha$.

The following useful properties of the permutations (50) may also be derived using the definition (49) or (50):

$$\text{(i)} \quad \hat{f}_i\,(P_g\vec{y}^1,\ldots,P_g\vec{y}^N) = (g\hat{f}_i) = (g\hat{f}_i)(\vec{y}^1,\ldots,\vec{y}^N). \tag{55}$$

This result states that the Eckart frame corresponding to the permuted position vectors $P_g\vec{y}^1,\ldots,P_g\vec{y}^N$ is the rotation g of the frame F corresponding to the position vectors $\vec{y}^1,\ldots,\vec{y}^N$.

$$\text{(ii)} \quad P_gQ_\mu = L_{g^{-1}}Q_\mu \ . \tag{56}$$

In this relation, P_gQ_μ is defined to be the result obtained from Q_μ by applying the permutation (50). Thus,

$$\begin{aligned} P_gQ_\mu &= \sum_{\alpha i} s^\alpha_{\mu+6,i}\,(g\hat{f}_i)\cdot(P_g\vec{\rho}^\alpha) \\ &= \sum_{\alpha i} s^\alpha_{\mu+6,i}\,\hat{f}_i\cdot(g^{-1}P_g\vec{\rho}^\alpha) = \sum_\alpha \vec{s}^\alpha_{\mu+6}\cdot L_{g^{-1}}\vec{\rho}^\alpha \\ &= L_{g^{-1}}Q_\mu \ . \end{aligned} \tag{57}$$

$$\text{(iii)} \quad P_{g'}(P_gQ_\mu) = P_{g'g}Q_\mu = (g^{-1}L_{g'^{-1}}g)(L_{g^{-1}}Q_\mu). \tag{58}$$

This relation may be proved by applying $P_{g'}$ to the first line of Eq. (57). Thus,

$$P_{g'}(P_g Q\ ) = \sum_{\alpha i} s^{\alpha}_{\mu+6,i}\ [g'(g\hat{f}_i)] \cdot [P_{g'}(P_g \vec{\rho}^{\alpha})] = P_{g'g} Q_{\mu}$$

$$= \sum_{\alpha} \vec{s}^{\alpha}_{\mu+6} \cdot g^{-1} g'^{-1} P_{g'}(P_g \vec{\rho}^{\alpha}) = \sum_{\alpha} \vec{s}^{\alpha}_{\mu+6} \cdot (g^{-1} L_{g'^{-1}} g)(L_{g^{-1}} \vec{\rho}^{\alpha})$$

$$= (g^{-1} L_{g'^{-1}} g)(L_{g^{-1}} Q_{\mu}) \quad .$$

We can now state the first important result on the permutation symmetry of the potential energy function $V(Q_1,\ldots,Q_{3N-6})$ of a rigid molecule: The potential energy function $V(Q_1,\ldots,Q_{3N-6})$ is invariant under the group $P(G)$:

$$V(P_g Q_1,\ldots,P_g Q_{3N-6}) = V(Q_1,\ldots,Q_{3N-6}) \ , \tag{59}$$

each $g \ \varepsilon \ G$.

Proof. Confer Eqs. (57) and (46).

The second important result relates to the representation of the group $P(G)$ as a semi-direct product group: The action of each permutation $P_g \ \varepsilon \ P(G)$ on the molecular coordinates $(F;Q)$, $F = (\hat{f}_1,\hat{f}_2,\hat{f}_3)$, $Q = (Q_1,\ldots,Q_{3N-6})$ can be represented by the ordered pair, $P_g = (g,\ L_{g^{-1}})$,

$$(g, L_{g^{-1}})(F;Q) = (gF;\ L_{g^{-1}} Q) \ , \tag{60}$$

where the multiplication rule for pairs is that of a semi-direct product:

$$(g',\ L_{g'^{-1}})(g, L_{g^{-1}}) = \left(g'g, (g^{-1} L_{g'^{-1}} g) L_{g^{-1}} \right) . \tag{61}$$

Proof. Confer Eqs. (55) and (58).

It is apparent from the preceding results that it is irrelevant as to which of the two groups $L(G)$ or $P(G)$ (each isomorphic to the point group G) we use to classify the internal coordinates according to their transformation properties under the irreducible representations of the point group G. In either case, it is the point group G itself which plays the fundamental role, and it is obtained from the static model of the molecule.

We have not yet encountered the feasible group. Let us now give its definition and discuss below its relation to the groups G, L(G), and P(G).

The feasible group F(G) of a rigid molecule is the union

$$F(G) = F_+(G) \cup F_-(G) , \tag{62}$$

where $F_+(g)$ is the subgroup of P(G) defined by

$$F_+(G) = \{P_g | g \in G \text{ and } \det R(g) = +1\} \tag{63}$$

and $F_-(G)$ is the set of operators defined by

$$F_-(G) = \{IP_g | g \in G \text{ and } \det R(g) = -1\} \tag{64}$$

in which I is the inversion operator on the space R^3.

The action of the group F(G) on the internal coordinates is the same as that of the group P(G), since each of the internal coordinates Q_μ is invariant under inversion. The action of the two groups on the Eckart frame is, however, different, since

$$\hat{f}_i(P_g\vec{y}^1,\ldots,P_g\vec{y}^N) = (g\hat{f}_i)(\vec{y}^1,\ldots,\vec{y}^N) ,$$

$$\hat{f}_i(IP_g\vec{y}^1,\ldots,IP_g\vec{y}^N) = -(g\hat{f}_i)(\vec{y}^1,\ldots,\vec{y}^N) . \tag{65}$$

Thus, the action of each element of the feasible group on the Eckart frame is always a pure rotation.

Summary. The isomorphic groups L(G), P(G), and F(G) have the actions on the molecular coordinates (F;Q) given by

	F	Q
L(G)	g	L_g
P(G)	g	$L_{g^{-1}}$
F(G)	g for det g = + 1 -g for det g = -1	$L_{g^{-1}}$

Critique. The Hamiltonian H of Eq. (31) [and (32)] is invariant under all pure rotations and inversion of R^3, and there is no apparent reason, in the case of rigid molecules, for introducing the feasible group since it occurs already as a subgroup of the group obtained by adjoining I to P(G). This same criticism applies, of course, to non-rigid molecules: The Hamiltonian for a model of a non-rigid molecule will be invariant under all rotation-inversions of the space R^3 as well as some subgroup of the group of permutations of the position vectors of identical particles [cf. Eq. (48)], the particular subgroup being fixed by the properties of the potential energy function which is chosen for the model, this choice itself being based on physical considerations. It is therefore difficult to see the advantages of introducing the feasible group in place of conventional approaches.

I conclude with some remarks about the point group. Does the concept of a "point group" generalize to non-rigid molecules ? Preliminary investigations (Ref. 6) indicate that it does for a class of such molecules, the key concepts being: (i) a definition of an appropriate body-fixed frame (or frames); and (ii) the definition of a group of transformations leaving the frame invariant.

REFERENCES

1. E. B. Wilson and J. B. Howard, The vibration-rotation energy levels of polyatomic molecules, J. Chem. Phys. 4 (1936), 260-268.

2. E. B. Wilson, J. C. Decius, and P. C. Cross, "Molecular Vibrations. The Theory of Infrared and Raman Vibrational Spectra," McGraw-Hill Book Co., New York, 1955.

3. J. D. Louck and H. W. Galbraith, Eckart vectors, Eckart frames, and polyatomic molecules, Rev. Mod. Phys. 48 (1976), 69-108.

4. J. D. Louck, Derivation of the molecular vibration-rotation Hamiltonian from the Schrödinger equation for the molecular model, J. Mol. Spectrosc. 68 (1977), 1-20.

5. C. Eckart, Some studies concerning rotating axes and polyatomic molecules, Phys. Rev. 47 (1935), 552-558.

6. H. W. Galbraith and J. D. Louck, Eckart frames and non-rigid molecules, Abstract RA1, Thirty Second Symposium on Molecular Spectroscopy, The Ohio State University, Columbus, Ohio, (1977).

Some suggestions concerning a geometric definition of the symmetry group of non-rigid molecules.

by Andreas W. M. Dress, Fakultät für Mathematik, Universität Bielefeld

0. Introduction

It is well known that the symmetry group of a rigid molecule can be considered either, geometrically, as a point group or, combinatorially, as a permutation or permutation-inversion group. For a non-rigid molecule H.C. Longuet-Higgins [1] has given a combinatorial definition of the molecular symmetry group by using his concept of f e a s i b l e operations - a concept, which was based on some earlier work by J. T. Hougen [2] and has found wide-ranging applications [3-5, for instance]. Still, for two reasons it seems desirable to have a simple geometric definition of the molecular symmetry group in the non-rigid case, as well. Firstly, one can hope for a better understanding of the concept of feasibility by basing it on explicit geometric considerations. Secondly, it is just the interplay of the geometric and the combinatorial definition which constitutes an essential part of the whole theory, as has been shown by work of J. D. Louck, for instance, [6].

In this paper such a simple geometric definition is being offered by representing the molecular symmetry group as a subgroup, not of the orthogonal group O_3 , but of a (cartesian or direct) product $O_3 \times O_3 \times \cdots \times O_3$ of as many copies of O_3 as there are rigid substructures in our molecule, or rather, since some symmetries may permute the rigid substructures among themselves, of the wreath-product of O_3 and Σ_n , the full symmetric group in n symbols, n being the number of rigid substructures. One possible application is a general definition of the spin double group of the molecular symmetry group which seems to be in agreement with the work of P. R. Bunker [7].

1. Spatial representations of molecules

In a surely idealizing, but often very practical way the spatial structure of a molecule can be described conveniently in terms of its skeleton S , i.e. a subset $S \subseteq \mathbb{E}^3$ [1)] of the three dimensional euclidean

[1)] For a formal definition of the mathematical symbols used in this text see the appendix space consisting of a finite number of points $P_{1,...,}P_n$ - representing the various nuclei - , a finite number of - generally - straight lines, connecting some of those points and representing the chemical bonds, and, may be, some additional features like circle arcs, connecting two straight lines, whose angle is supposed to be fixed, or little balls of varying diameter instead of points so that nuclei of different species may be distinguished geometrically, or just symbols (C,H,B,F,...) attached to the points specifying the type of the nucleus at that position-according to chemical nomenclature. In a formal way, the latter can be described as a (settheoretic) map f from the finite set $\mathcal{P}=\{P_{1,...,}P_N\}$ into the finite set $\mathcal{M}$ of chemical symbols for the various possible types of nuclei. One may also add further geometrical features according to whatever may seem to be practical or desirable.

Anyway: what one ends up with is a subset $S \subseteq E^3$, the skeleton, a specified subset $\mathcal{P} = \{P_{1,...,}P_N\} \subseteq S$, the set of positions of nuclei and, may be, a set-theoretic map $f : \mathcal{P} \rightarrow \mathcal{M}$, $\mathcal{M}$ being some appropriate finite set, specifying the types of nuclei being involved. The first formal question that arises with respect to such a desription of the spatial structure of a molecule is, of course, the following : under what conditions do two such structures $(S,\mathcal{P},f)$ and $(S',\mathcal{P}',f')$ describe the "same" molecule?

Since we do not have the molecule, as a given entity on the one hand, and its spatial representations, on the other hand, which then could be compared somehow with the given molecule to answer this question, but since, instead, the spatial structure of the molecule can only be articulated in terms of some spatial representation, the only rigorous and formally satisfying way to answer such a question, known so far among mathematicians, is by defining an equivalence relation [1)] on such structures $(S,\mathcal{P},f)$, $(S',\mathcal{P}',f')$,..., such that $(S,\mathcal{P},f)$ is equivalent to $(S',\mathcal{P}',f')$ with respect to this relation if and only if they are meant to represent the same molecule - which itself, by this definition, is nothing else than just the full equivalence class, consisting of all of its representations $(S,\mathcal{P},f)$.

Remark 1: Similarly, the number 3, for instance, does not exist - mathematically - as a separate entity outside the world of counting but only through its representations by sets of three elements (or by some constructive device).

Of course, the definition of such an equivalence relation is not a purely geometric task, but relies heavily on the structure of the molecule as understood by the chemists, i.e. chemical knowledge is not only used, in a static way, to construct the representing spatial structures (S,P,f), but also, in a more dynamic way, to define the relation of equivalence for all such structures which describe the same molecule.

Geometry is usefull to offer various possible such definitions and to unfold their consequences so that they can be tested against the knowledge and the intuitive concept of the chemist. This is precisely the (surely very restricted) purpose of this note.

2. Rigid molecules.

The first possible definition for equivalence of spatial representations $(S,P,f),(S',P',f'),\ldots$ of a molecule is, of course, proper congruence: two representations are said to be equivalent (i.e. to represent the same molecule) if and only if they are properly congruent, i.e. if and only if there exists a proper, distant preserving map $\alpha : \mathbb{E}^3 \to \mathbb{E}^3$ in $O^+(\mathbb{E}^3)$[1] with $\alpha(S) = S'$, $\alpha(P) = P'$ and $f(P_i) = f'(\alpha(P_i))$.
In this case the molecule represented by the equivalence class $(S,P,f),(S',P',f'),\ldots$ will be said to be a rigid molecule (- represented in its equilibrium state, i.e. neglecting vibrations, of course. We will touch those at the end of this paper).

Remark 2: Geometrically, one could, of course, use the notion of congruence instead of proper congruence to define an equivalence relation, thus working with maps α in $O(\mathbb{E}^3)$ instead of $O^+(\mathbb{E}^3)$. This way one would identify molecules and their enantiomers, thus neglecting the phenomenon of chirality. Though this may make sense at some occasions, we do not want to do this here and therefore use proper congruence to define rigid molecules and their equivalence relation. Only for achiral molecules both definitions agree.

A spatial movement of such a rigid molecule in its equilibrium state, moving from a spatial position, described by (S,P,f) at time t_0 to a position (S',P',f') at time t_1 can now be described as a continuous map.

$$A : [t_0,t_1] \to O^+(\mathbb{E}^3)$$

From the time intervall $[t_0,t_1]$ into the group $O^+(\mathbb{E}^3)$ of proper, distant preserving maps from $\mathbb{E}^3$ into $\mathbb{E}^3$ such that $A(t_0) = Id_{\mathbb{E}^3}$ and $A(t_1) = \alpha$ with $\alpha(S) = S'$, $\alpha(P) = P'$, $f(P_i) = f'(\alpha(P_i))$ $(i = 1,\dots, n)$. (Again we neglect vibrations at this stage of our set up of definitions).

The symmetry group of such a rigid molecule, represented by (S,P,f) is defined, in a first step, to be the group $G = G_S = G(S,P,f)$ of all maps $\alpha \in O(\mathbb{E}^3)$ (and not only $O^+(\mathbb{E}^3)$!) with $\alpha(S) = S$, $\alpha(P) = P$, $f(P_i) = f(\alpha(P_i))$:

$$G_S = G_{(S,P,f)} = \{\alpha \in O(\mathbb{E}^3) \mid \alpha(S) = S; \alpha(P) = P; f(P_i) = f(\alpha(P_i)), i = 1,\dots,n\}.$$

Its acts on the finite set P, respecting the various species, and thus it can be considered not only as a subgroup of $O(\mathbb{E}^3)$, i.e. as a point group (as which it occurs by its very definition), but also as well as a sub-group of the symmetric group Σ_N — or rather its subgroup $\Sigma_{N_1} \times \dots \times \Sigma_{N_r}$, N_i being the number of nuclei of the i-th species $(N = N_1 + N_2 + \dots N_r)$ -, i.e. as a permutation group.

After some thought, one might realize that our definition of the symmetry group depends not only the molecule, but on the specified representation (S,P,f). Another representation (S',P',f') leads to another symmetry group $G_S = G_{(S',P',f')}$, of course. But if for some $\alpha \in O^+(\mathbb{E}^3)$ one has $\alpha(S) = S'$, $\alpha(P) = P'$, $f(P_i) = f'(\alpha(P_i))$ $(i = 1,2,\dots, n)$, then obviously $G_{S'} = \alpha\, G_S\, \alpha^{-1}$. So t h e symmetry group of the rigid molecule represented by (S,P,f) is not so much the group G_S, but rather the class $\{\alpha\, G_S\, \alpha^{-1} \mid \alpha \in O^+(\mathbb{E}^3)\}$ of conjugate subgroups of $O(\mathbb{E}^3)$. The same way, another numbering of the nuclei will lead to another, but conjugate subgroup of Σ_N.

Remark: Of course, the "abstract" group associated with G_S , i.e. its abstract isomorphism class in the category of groups, is also an invariant of the molecule itself and not dependend on its various spatial representations (conjugate subgroups are allways isomorphic !). But this is a much wider and therefore less specific class. In other words, the symmetry group of a rigid molecule is not only an abstract group G , but it is such a group G endowed with an action[1] of G on $\mathbb{E}^3$ by distant preserving maps and an action of G on the finite set of nuclei by permutations. Both of these actions are related to each other in the way described above.

3 Non-rigid molecules with rigid substructures

Let us now look at non-rigid molecules. As pointed out in § 1, their "non-rigidity" has to be reflected in form of an adequate definition of equivalence of their various spatial representations. As such a definition I would like to propose the following one: At first one defines a "rigid partition" of (S,P,f) or just $S \subseteq \mathbb{E}^3$ to be an indexed set of subsets $S_1,\ldots, S_n \subseteq S$ such that $\bigcup_{j=1}^{n} S_j = S$.

Each S_i is supposed to consist firstly of those nuclei (i.e. points P_i) whose relative positions with respect to each other are supposed to be fixed and rigid - up to vibrations - , while the non-rigid molecule tumbles through space, secondly the straight lines connecting them and thirdly of what ever geometric features might be involved with those nuclei in the build up of the skeleton of our molecule.

How to fix the S_i for a given molecule is basically a matter of the chemist - though some mathematical comments on this problem are possible. Those have been discussed allready with A. Dreiding and may be the subject

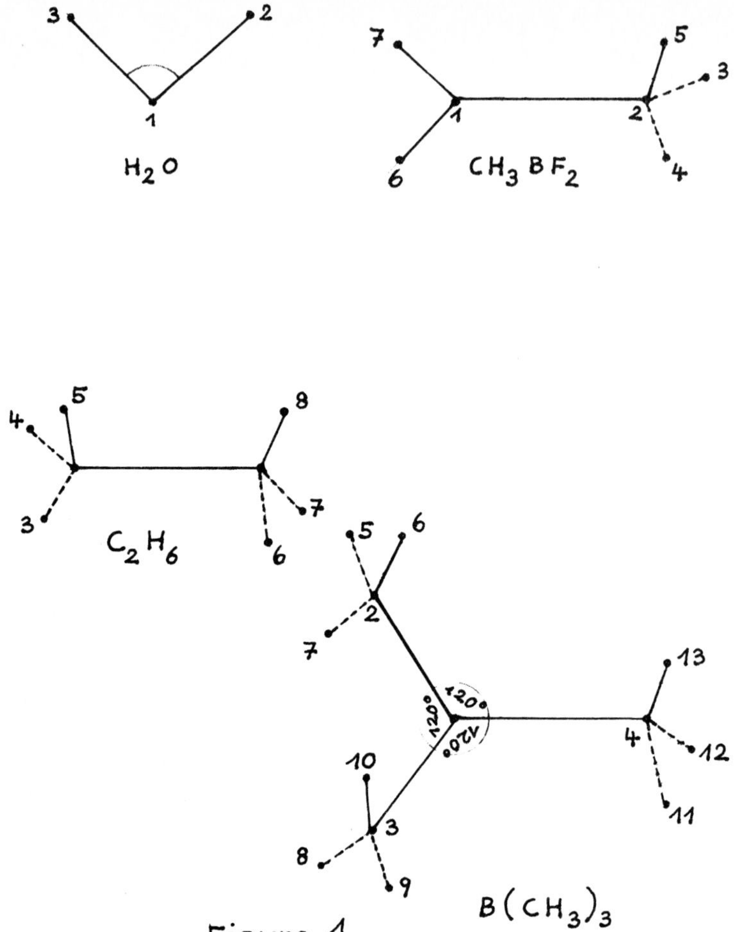

Figure 1

on some future, joined work.

In the examples, given in fig. 1, an obvious choice for $C\,H_3\,B\,F_2$ would be (with $\overline{P_i\,P_j}$ denoting the straight line from P_i to P_j):

$$S_1 = \overline{P_1\,P_2} \cup \overline{P_1\,P_6} \cup \overline{P_1\,P_7}$$

and

$$S_2 = \overline{P_1\,P_2} \cup \overline{P_2\,P_3} \cup \overline{P_2\,P_4} \cup \overline{P_2\,P_5}$$

For $C_2\,H_6$ it is similarly

$$S_1 = \overline{P_1\,P_2} \cup \overline{P_1\,P_3} \cup \overline{P_1\,P_4} \cup \overline{P_1\,P_5}$$

and

$$S_2 = \overline{P_1\,P_2} \cup \overline{P_2\,P_6} \cup \overline{P_2\,P_7} \cup \overline{P_2\,P_8}$$

For $B\,(C\,H_3)_3$ it is

$$S_1 = \overline{P_1\,P_2} \cup \overline{P_1\,P_3} \cup \overline{P_1\,P_4}\,,$$

$$S_2 = \overline{P_1\,P_2} \cup \overline{P_2\,P_5} \cup \overline{P_2\,P_6} \cup \overline{P_2\,P_7}\,,$$

$$S_3 = \overline{P_1\,P_3} \cup \overline{P_3\,P_8} \cup \overline{P_3\,P_9} \cup \overline{P_3\,P_{10}}$$

$$S_4 = \overline{P_1\,P_4} \cup \overline{P_4\,P_{11}} \cup \overline{P_4\,P_{12}} \cup \overline{P_4\,P_{13}}\;.$$

Now let $S'_1, S'_2, \cdots S'_m$ be a rigid partition of S'. We define $(S,P,f;\,S_1,\cdots,S_n)$ to be equivalent to $(S',P',f';\,S_1',\cdots,S_m')$ if $n = m$ and if there exist mappings $\alpha_1,\cdots,\alpha_n \in 0^+(\mathbb{E}^3)$ and $\pi \in \Sigma_n$

(1) $\alpha_j(S_j) = S_{\pi(j)}, \alpha_j(S_j \cap' P) = S_{\pi(j)} \cap P'$,
$f'(\alpha_j(P_i)) = f(P_i)$ for all $P_i \in S_j \cap P$;

(2) $\alpha_j\,(S_j \cap S_k) = \alpha_k\,(S_j \cap S_k) = S'_{\pi(j)} \cap S'_{\pi(k)}$

(3) $\alpha_j\,(P) = \alpha_k\,(P)$ for all $P \in S_j \cap S_k$; $j,k = 1,\cdots,n$.

Consequently a spatial movement of $(S,P,f;\,S_1,\cdots,S_n)$ can be de-

scribed as a continuous map A from the time intervall $[t_0,t_1]$ into the set $A_{(S;S_1,\cdots,S_n)}$ of all n-tupels $(\alpha_1,\cdots,\alpha_n) \in O^+(\mathbb{E}^3) \times\cdots\times O^+(\mathbb{E}^3)$ with $\alpha_j(S_j \cap S_k) = \alpha_k(S_j \cap S_k) = \alpha_j(S_j) \cap \alpha_k(S_k)$ and $\alpha_j(P) = a_k(P)$ for all $P \in S_j \cap S_k$; $j,k = 1,\cdots,n$ starting with $A(t_0) = Id_{\mathbb{E}^3}\times\cdots\times Id_{\mathbb{E}^3}$
The set $A_{(S;\, S_1,\cdots,S_n)}$ may be called the set of "admissible" mappings of $(S;\, S_1,\cdots,\, S_n)$.

The symmetry group $G = G_S = G_{(S,P,f;\, S_1,\cdots,\, S_n)}$ is defined, accordingly, as the set of all $(n + 1)$-tupels $(\pi;\, \alpha_1,\cdots,\alpha_n)$ with $\pi \in \Sigma_n;\, \alpha_1,\ldots,\, \alpha_n \in O(\mathbb{E}^3)$ (again not only $O^+(\mathbb{E}^3)$!) with $\det \alpha_1 = \det \alpha_2 =,\ldots,= \det \alpha_n$ satisfying the conditions (1)-(3) above with $(S',P',f';\, S_1',\cdots,\, S_n') = (S,P,f;\, S_1,S_2,\cdots,\, S_n)$.

The composition of two such $(n+1)$-tupels $(\pi;\, \alpha_1,\cdots,\alpha_n)$ and $(\pi';\, \alpha_1',\cdots,\alpha_n')$ is defined - as in the case of wreath-products - by $(\pi;\, \alpha_1,\cdots,\, \alpha_n) \circ (\pi';\, \alpha_1',\cdots,\, \alpha_n') = (\pi \cdot \pi';\alpha_{\pi'(1)}\cdot\alpha_1',\cdots,\alpha_{\pi'(n)}\cdot\alpha_n')$.
So one has, of course, $(\pi;\alpha_1,\cdots,\alpha_n) \circ (\pi';\alpha_1',\cdots,\alpha_n') \in G_S$.

The p r o p e r symmetry group is
$G^+ = G_S^{\,+} = G^+_{(S,\;,\;;S_1,\cdots,S_n)} = \{(\pi;\alpha_1,\cdots,\alpha_n)\in G_S \mid \det \alpha_1 = \ldots = \det \alpha_n = 1\}$.
$(S,P,f;\, S_1,\cdots,\, S_n)$ will be defined to be achiral, if $G_S^{\,+} \neq G_S$ (thus $(G_S : G_S^{\,+}) = 2$), and chiral, if $G_S^{\,+} = G_S$. Similarly, one may define an enantiomer of $(S,P,f;\, S_1,\cdots,\, S_n)$ to be any structure $(S',P',f';\, S_1',\cdots,\, S_n')$ for which there exist $\alpha_1,\cdots,\alpha_n \in O^-(\mathbb{E}^3) = O(E^3) \setminus O^+(E^3)$ and $\pi \in \Sigma_n$ satisfying the relations (1)-(3), above.

Thus, $(S,P,f;\ S_1,\dots,S_n)$ is chiral if and only if it is not equivalent to its enantiomers.

The spin double group

$Sp_S = Sp_{(S,P,f;\ S_1,\dots,S_n)}$ of $(S,P,f;\ S_1,\dots,S_n)$ can be defined as the group consisting of all (n+1)-tupels $(\pi;\ \beta_1,\dots,\beta_n)$ with $\pi \in \Sigma_n$; $\beta_1,\dots,\beta_n \in \mathrm{Spin}(\mathbb{E}^3)^1$ such that $(\pi;\ \bar\beta_1,\dots,\bar\beta_n) \in G_S$ modulo the central normal subgroup of all (n+1)-tupels $(\mathrm{Id}_{\{1,\dots,n\}};\ \varepsilon_1,\dots,\varepsilon_n)$ with $\bar\varepsilon_j = \mathrm{Id}_{\mathbb{E}^3}$ and $\varepsilon_1 \cdot \varepsilon_2 \cdot \dots \cdot \varepsilon_n = 1_{\mathrm{Spin}(\mathbb{E}^3)}$.
Thus $Sp_S \to G_S : (\pi;\ \beta_1,\dots,\beta_n) \to (\pi;\ \bar\beta_1,\dots,\bar\beta_n)$ is a welldefined surjective group homomorphism, whose kernel consists of exactly two elements.

Each element $(\pi;\ \alpha_1,\dots,\alpha_n) \in G_S$ acts moreover as a permutation on the set P of nuclei, permuting identical nuclei among themselves: for $P_i \in P$ with $f(P_i) = X \in M$ one has $P_i \in S_j$ for some $j \in \{1,\dots,n\}$ and thus $\alpha_j(P_i) \in \alpha_j(P \cap S_j) = P \cap S_{\pi(j)} \subseteq P$, i.e. $\alpha_j(P_i) = P_{i'}$ for some $i' \in \{1,\dots,N\}$ with $f(P_{i'}) = f(\alpha_j(P_i)) = f(P_i) = X$. Thus by associating to any element $(\pi;\ \alpha_1,\dots,\alpha_n) \in G_S$ either the permutation $\overline{(\pi;\ \alpha_1,\dots,\alpha_n)} : i \mapsto i'$ of $\{1,\dots,N\}$ onto itself in case $\det \alpha_1 = \dots = \det \alpha_n = 1$ and the permutation inversion $\overline{(\pi;\ \alpha_1,\dots,\alpha_n)}^*$: $i \to i'$ in the case $\det \alpha_1 = \dots = \det \alpha_n = -1$, we get a representation of G_S as a permutation inversion group.

In the examples from fig. 1 the permutation-inversion-groups associated with the rigid partitions given above are, of course, precisely those which have been listed by Longuet-Higgings. In the case of $B(CH_3)_3$, for instance, the first subset S_1 will allways be mapped onto itself; the subgroup G_1, fixing S_1, is a subgroup of index 6 (the factor group G/G_1 acts on $\{P_2,P_3,P_4\}$ as the full permutation group); in G_1 we have $G_1 \cap G_S^+$ as a subgroup of index 2; $G_1 \cap G_S^+$ acts independently by even permutations on $\{P_5,P_6,P_7\}$, $\{P_8,P_9,P_{10}\}$, $\{P_{11},P_{12},P_{13}\}$ and thus has order $3 \cdot 3 \cdot 3 = 27$. Thus the whole group has order $6 \cdot 2 \cdot 27 = 324$ and one may realize that - though it is not a direct product of much simpler subgroups, as has been pointed out by Longuet-Higgins allready [1,p. 453], - it is easily decomposed into its group-theoretic constituents by using its geometric definition.

How does the group act on the space of vibrations? If we allow each point P_i to vibrate freely inside a small sphere around P_i, the space of vibrations will have $3N = 3 \cdot 13 = 39$ degrees of freedom. Though this will be too much, since it counts admissible mappings close to $Id_{\mathbb{E}^3} \times \cdots \times Id_{\mathbb{E}^3}$ as vibration as well (thus the real degree of freedom of the space of vibrations of our molecule is $39 - 12 = 27$, 12 being the dimension of the space of admissible mappings), it still contains the space of "proper" vibrations as a subspace and thus it is enough to describe the action of G_S on the whole space of dimension $3N = 39$. (To identify the subspace of "proper" vibrations, the theory of the Eckart Frame, as developed by J. D. Louck and H. W. Galbraith, appears to be the appropriate tool).

To describe the action of G_S on the whole space of dimension 39, we split this space into a product

$R_1^3 \times (R_2^3 \times R_3^3 \times R_4^3) \times (R_5^3 \times\cdots\times R_7^3) \times\cdots\times (R_{11}^3 \times\cdots\times R_{13}^3)$.

An element $(\pi; \alpha_1, \cdots, \alpha_4) \in G_s$ will then act on a vector

$(x_1; x_2, x_3, x_4; x_5, x_6, x_7 ;\cdots; x_{11}, x_{12}, x_{13})$

by

$(\pi; \alpha_1, \alpha_2, \alpha_3, \alpha_4) \cdot (x_1 ,\cdots, x_{13}) = (y_1 ,\cdots, y_{13})$

with

$$\alpha_1 (x_1) = y_1$$

....

$$\alpha_2 (x_2) = y_{\pi(2)}$$

$$\alpha_3 (x_3) = y_{\pi(3)}$$

$$\alpha_4 (x_4) = y_{\pi(4)}$$

....

$$\alpha_2 (x_5) = y_{5'}, \text{ if } \alpha_2 (P_5) = P_{5'}$$

$$\alpha_2 (x_6) = y_{6'}, \text{ if } \alpha_2 (P_6) = P_{6'}$$

$$\alpha_2 (x_7) = y_{7'}, \text{ if } \alpha_2 (P_7) = P_{7'}$$

....

$$\alpha_3 (x_8) = y_{8'} \text{ if } \alpha_3 (P_8) = P_{8'}$$

$$\alpha_3 (x_9) = y_{9'} \text{ if } \alpha_3 (P_9) = P_{9'}$$

$$\alpha_3 (x_{10}) = y_{10'} \text{ if } \alpha_3 (P_{10}) = P_{10'}$$

....

$$\alpha_4 (x_{11}) = y_{11'} \text{ if } \alpha_4 (P_{11}) = P_{11'}$$

$$\alpha_4 (x_{12}) = y_{12'} \text{ if } \alpha_4 (P_{12}) = P_{12'}$$

$$\alpha_4 (x_{13}) = y_{13'} \text{ if } \alpha_4 (P_{13}) = P_{13'} .$$

In general, one will have $y_{i'} = \alpha_j (x_i)$ if $P_i \in S_j, \alpha_j(P_i) = P_{i'}$ and if $x_i \in \mathbb{R}^3$ is contained in the vectorspace, spanned by all $\overline{P_i P_k}$ with $\overline{P_i P_k} \subseteq S_j$. - If there is any interest in this definition of the action of G_S on the space of vibrations by chemists, details may be worked out in another paper. In any case, the definition shows, that the geometric definition is needed together with the combinatorial definition (i.e. the permutation $P_i \rightarrow P_{i'}$) to describe the action of G_S on the space of vibrations, thus corroborating the arguments, put forward by J.D.Louck/H.W.Galbraith.

Finally, if $(S, P, f; S_1 ,\cdots, S_n)$ and $(S', P', f'; S' ,\cdots, S_n')$ are equivalent, then - as in the rigid case - the associated groups are conjugate and are mapped into one another by means of the "admissible" (n+1)-tupel $(\pi; \alpha_1 ,\cdots, \alpha_n)$ which transforms $(S, P, f; S_1' ,\cdots, S_n')$ into $(S', P', f'; S_1' ,\cdots, S_n')$. It are actually those classes of conjugate subgroups of the wreath product of $O(\mathbb{E}^3)$ and Σ_n , which are t h e symmetry group of the molecule in question - again in complete analogy to the rigid case.

4 Concluding remarks

The above definitions are put forward only as a suggestion. Almost no formal mathematical consequences have been developed, though it was tempting to indulge in such a business. I have refrained, not only to keep this paper readible, but also, since I believe that before doing so it is first up to the chemists to decide upon wether the whole idea is worthwhile to be developed in detail or not. So far it is only the believe into the usefullness of mathematically structured models in general, which makes me hope, that the above suggestions might be chemically relevant.

References

1. Longuet-Higgins, H. C., 1963, Molec. Phys., 6, 445.

2. Hougen, J. T., 1962, J. chem. Phys., 37, 1433; 1963; 38, 358.

3. Stone, A. J. 1964, J. chem Phys., 41, 1568.

4. Woodman, C. M., 1970, Molec. Phys., 19, 753.

5. Hougen, J. T., 1975, "Catalog of Explicit Symmetry Operations..." in MTP International Review of Science: Physical Chemistry Series, ed. by D. A. Ramsay.

6. Louck, J. D., these proceedings.

7. Bunker, P. R., these proceedings and "Molecular Symmetry and Spectroscopy", Academic Press, 1978.

Appendix

An equivalence relation on a set M is a two-valued relation aRb, generally written in the form $a \sim b$, $(a,b \in M)$, such that:

(i) $a \sim a$ for all $a \in M$

(ii) $a \sim b$ and $b \sim c \Rightarrow a \sim c$ for all $a,b,c \in M$

(iii) $a \sim b \Rightarrow b \sim a$ for all $a,b, \in M$.

The equivalence relations are used to identify mathematical objects with respect to whatever aspect one might have in mind.

An action of a group G on a set (or space) M is a map $G \times M \to M : (g,m) \quad g \cdot m$, such that

(i) $1_G \cdot m = m$ (1_G the unit element in G) ,

(ii) $g \cdot (h \cdot m) = (g \cdot h) \cdot m \ (g,h, \in M)$

(and - in case M is a real or complex vectorspace -

(iii) $g \cdot (m + n) = g \cdot n$, $g \cdot (c \cdot m) = c \cdot (g \cdot m)$

for $g \in G$; $m,n \in M$; $c \in \mathbb{R}$ (or $\mathbb{C}$)).

Thus any $g \in G$ defines a permutation $\overline{g}$ (linear automorphism) of $M : \overline{g}\ (m) = g \cdot m \quad (m \in M)$, $G \to \mathrm{Aut}\ (M) : g \mapsto \overline{g}$ is a group homomorphism.

In particular the full symmetric group Σ_n acts by permutations on $\{1,\ldots, n\}$, the above homomorphism $\Sigma_n \to \mathrm{Aut}\ (1,\ldots, n\}) \ (= \Sigma_n)$ is just the identity.

The 3-dimensional euclidean space $\mathbb{E}^3$ is defined to be a set $\mathbb{E}^3$ together with an action of the additive group $\mathbb{R}^3$ (the 3-dimensional real vectorspace of real 3-tupels (x_1, x_2, x_3) , $x_i \in \mathbb{R}$) on $\mathbb{E}^3$ such that for any $P, Q \in \mathbb{E}^3$ there exists precisely one vector $(x_1, x_2, x_3) \in \mathbb{R}^3$ with $(x_1, x_2, x_3) \cdot P = Q$. The distance $d(P,Q)$ of P and Q is in this case defined to be $\sqrt{x_1^2 + x_2^2 + x_3^2}$. A distance preserving map $\alpha: \mathbb{E}^3 \to \mathbb{E}^3$ is a set-theoretic map with $d(\alpha(P), \alpha(Q)) = d(P,Q)$ for all $P,Q \in \mathbb{E}^3$. It is well known, that any such α can be represented in the form of a product of a translation and a (proper or unproper) rotation, if and only if it has a fixed point P $(\alpha(P) = P)$. All such distant preserving maps of $\mathbb{E}^3$ onto $\mathbb{E}^3$ form a group $O(\mathbb{E}^3)$, which can be represented as the semidirect product of the standard orthogonal group and $\mathbb{R}^3$; one has a canonical map $\det : O(\mathbb{E}^3) \to \pm 1$, mapping proper, orientation preserving maps onto $+1$ and orientation reversing maps onto -1 . One puts $O^+(\mathbb{E}^3) = \{\alpha \in O(\mathbb{E}^3) \mid \det \alpha = +1\}$ and $O^-(\mathbb{E}^3) = \{\alpha \in O(\mathbb{E}^3) \mid \det \alpha = -1\}$ $= O(\mathbb{E}^3) \setminus O^+(\mathbb{E}^3)$. The unit element in $O(\mathbb{E}^3)$ is, of course, the identity map $Id_{\mathbb{E}_3}$ on $\mathbb{E}^3$.

There exists a universal (two-fold) covering group of $O(\mathbb{E}^3)$, called spin $(\mathbb{E}^3)$, which maps surjectively onto $O(\mathbb{E}^3)$. For $\beta \in \mathrm{Spin}\,(\mathbb{E}^3)$ the canonical image of β in $O(\mathbb{E}^3)$ is denoted by $\overline{\beta}$.

SYMMETRY AND THERMODYNAMICS FROM STRUCTURED MOLECULES TO LIQUID DROPS

R. Stephen Berry
Department of Chemistry and the James Franck Institute
The University of Chicago
Chicago, Illinois 60637, U.S.A.

Abstract

By idealizing both limits, one can construct a correlation diagram for the quantum states of a system of N identical particles whose extremes are a molecule-like polyhedron with separable vibrations and rigid-body rotations, and a fluid cluster. The conceptual approach offers a way to analyze vibration-rotation spectra of nonrigid molecules, and an artifice to interpret the melting and nucleation of small clusters. The approach leads to the use of a conditional probability density in r_1 and θ_{12} for description of two particles bound to a single center (where r_2 is fixed), as one probe of the tendency of such a system to adopt a polyhedral structure. A single dimensionless parameter based on the ratio of two energy levels provides a characterization of the degree of nonrigidity of a cluster.

Introduction

This report is a summary of the results achieved thus far and something of a preview of work nearing completion that has been carried out by our group at the University of Chicago: Michael Kellman,* Paul Rehmus, Francois Amar and myself. The focus of the work is the phenomenological description of clusters, especially nonrigid clusters of identical particles. We have addressed three levels of problems. One is the description of the energy levels and vibration-rotation spectra of such clusters and how they relate to the levels and spectra of rigid molecules. The second is the description of spatial correlation in small clusters. The third is the analysis of the density of states and the thermodynamic properties of nonrigid clusters.

Molecules and solids play a role unique among the structures we find in nature. All the other structures we observe, from nucleons and nuclei to galaxies and clusters of galaxies, are fluid, nonrigid systems. Only

*Now at the Institute for Theoretical Studies, The University of Oregon, Eugen, Oregon, U.S.A.

molecules and solids have "permanent" structures in which the component parts occupy easily-specified sites for times long compared with the times of our observations. Because they are so rigid, most free molecules can be described as rigid rotors that undergo small-amplitude vibrations, and their vibrational motion is nearly separable from their rotational motion.

The rigidity of molecules creates a fascinating apparent paradox: (1) identical particles are supposed to be indistinguishable, but chemists continually do experiments that distinguish one hydrogen atom from another in the same molecule, because they occupy chemically inequivalent sites. The very existence of chemically inequivalent sites, the result of molecular rigidity, seems to contradict a fundamental postulate of quantum mechanics. The explanation of the "paradox" lies of course in the time scale of our observation. The requirement of the basic postulate of indistinguishability of identical particles only applies to stationary states. If we observe the system in a nonstationary state, it may be quite possible to distinguish one proton from another. Indistinguishability is only accomplished when the identical particles can exchange sites with one another. The stationary states of a system in which exchanges or rearrangements can occur are separated in energy by intervals h whose frequencies are characteristic of the rates of exchange. In atoms, the electrons look equivalent to us because their exchange rates or exchange integrals are of order 10^{13} - 10^{14} sec^{-1}. The protons in the middle of a propane molecule, H_3C-CH_2-CH_3, are distinguishable to a nuclear resonance spectrometer because the rate at which these protons exchange with those on either CH_3 group is probably less than once per century. The time scales for exchanges of atoms can range from fractions of picoseconds to eons. It is worth noting that the "paradox of distinguishability" continues to attract attention in one guise or another.(2)

Sometimes--often enough to be not only interesting but even important--Nature gives a molecule a means to make its identical atoms equivalent at a rate that falls in the range of our ability to observe. Examples are the inversion of ammonia, the pseudorotation of the 5-coordinate compounds of phosphorus such as PF_5, the positive ions of many hydrocarbons and the triatomic alkali molecules such as Li_3 and Na_3. Many van der Waals molecules are probably able to rearrange on the time scale between a few picoseconds and a few milliseconds; perhaps clusters of more than three alkali atoms are also able to rearrange rapidly. Let us call such molecules nonrigid or floppy.

II. Symmetry of Nonrigid Clusters

To introduce symmetry into the problem of nonrigid molecules of N identical atoms, we examine the potential energy hypersurfaces, $V(\vec{R}_1,\ldots\vec{R}_N)$ that govern their nuclear motion. If there are two or more equivalent ways to assign nuclei to sites, the potential surface has a corresponding number of equivalent minima. If the molecule is rigid, in the sense that the system cannot exchange particles and thereby move from one well to another equivalent well on the time scale of our observations, then the energy levels exhibit at least an apparent degeneracy equal to the number of equivalent wells. (There may, of course, be additional degeneracy.) If the wells are connected and the particles can exchange observably among them, then the degeneracy associated with the equivalent minima is split. When our observations are too fast to resolve the stationary states, we catch the system in a localized, nonstationary state made of the coherent localized superposition of stationary states. Such a state is a solitary wave, an oldfashioned standing solitary wave.

If the system is rigid, then its Hamiltonian has the symmetry of the point group of the rigid polyhedron formed by the nuclei, times the orthogonal group O(3) arising from the rotational invariance of the Hamiltonian. If identical atoms can rearrange, the symmetry of the system must be higher than the point group; if all the N identical particles can exchange with one another, the symmetry group must include the permutation group S_N. The expansion of the symmetry of a molecule from the point group to something larger was explored by Longuet-Higgins (3), Watson (4), Hougen (5) and Bunker (6). If a molecule has a path available that permits it to pass from one potential well to another, one can define an operation R corresponding to that process, and construct the products of R with each element of the point group P of the molecule in a single well. This way, R acts as a generator of a new set of elements RP. This is obviously one coset of a group generated by R and P, a group called the Q-group. (3)

The relation between the symmetry of the Q-group and the rotation-vibration spectrum of a nonrigid molecule was worked out for nontrivial cases first by Dalton (7). The approach provided clear rules for how to observe the splittings due to exchanges between potential minima, in the perturbation limit where the large-amplitude motions that accomplish particle exchanges are slow relative to rotational periods. That is, Dalton's solution is valid when the energy level splittings due to nonrigidity are smaller than the separations of the rotational states. We shall return to this point shortly.

During the 196o's, a controversy arose, especially among inorganic chemists, concerning what should be the main matter of concern with regard to nonrigid molecules. One school (8, 9) held that the topology of the potential surface and its feasible (observable) paths of rearrangement are what matters; the other school (1o) argued that one had to understand the details of the path that the molecule follows from one well to another, which can be reworded to say that one must understand the geometry of the potential surface, especially in the region above the energy level from which the molecule starts. We, with the sage perspective of physical chemists, physicists and mathematicians, recognize immediately that the important thing in such a problem is a knowledge of where the quantum states are. The starting point is the rigid molecule with its degenerate levels; the topology of the surface tells us the character of the level splitting, and the geometry of the surface tells us its magnitude. We need to know both in order to understand the real issue.

Expressing the problem in terms of energy levels takes us to the next stage of our analysis. We think of the rigid molecule as a limiting case. Is there any other limiting case? There is indeed: the completely nonrigid, uncorrelated cluster. This limiting case appears in several forms but they all have one common representation, the Hartree-Fock self-consistent field, with each particle free of all the others except that it moves in their average field.

Deviations from this uncorrelated limit are better understood than the deviations from the rigid limit. Here we know how to construct a representation in terms of configurations defined over sets of one-particle states, and then introduce correlation by superposing configurations. The representation of electron correlation, for example, is usually accomplished this way in all but the simplest systems, where interelectronic distances can be used explicitly.

III. Construction of the Correlation Diagram

Now we come to the central question from which the rest of this discussion follows: how can we construct an energy-level correlation diagram to connect the rigid limit with the nonrigid, uncorrelated limit? To do this, we must find ways, approximate though they may be, to represent the energy levels of the two limiting cases, determine what are the quantum numbers of the states at both the limits and in the region between, and then connect the limiting levels that have the same quantum numbers throughout, according to the adiabatic theorem (11). The aspect of the problem that goes beyond well-known mathematical techniques is the selection of idealized limiting cases of sufficiently high symmetry to allow us to construct approximate spectral distributions and of sufficient plausibility to let us relate the idealized distributions to those of real molecules.

The nonrigid limit is easily chosen as the one long used in nuclear physics (12), a set of N identical particles interacting with pairwise harmonic forces with zero equilibrium distance. The Hamiltonian for this system,

$$H = \sum_{i=1}^{N} \frac{\vec{P}_i}{2m} + \frac{1}{2} K \sum_{i\,j}^{N} (\vec{r}_i - \vec{r}_j)^2, \tag{1}$$

can be transformed to a center-of-mass system and the motion of the center of mass can be dropped; successive orthogonalization and mormalization of the pair coordinates $(\vec{r}_i - \vec{r}_j)$ generates a set of normal coordinates ρ_j (Jacobi vectors) in terms of which the internal Hamiltonian becomes

$$H_{int} = \frac{1}{2m} \sum_{k}^{N-1} \vec{p}_{\rho_k}^{\,2} + \frac{1}{2}(NK) \sum_{k}^{N-1} \vec{\rho}_k^{\,2} \tag{2}$$

This is the Hamiltonian for a 3(N-1)-dimensional isotropic harmonic oscillator well-known to be invariant under the unitary group U(3N-3)(13).The eigenstates corresponding to ν quanta are the basis of the νth totally symmetric irreducible representation of this group--totally symmetric because the quanta are bosons.

The representations can be decomposed and classified according to parity, total angular momentum and permutation symmetry. To do this, we use the subgroup chain (14)

$$\begin{array}{l} U(3N-3) \equiv U(1) \times SU(3N-3) \supset \underbrace{U(1) \times SU(N-1)} \times SU(3) \\ \qquad\qquad\qquad\qquad\qquad\qquad\qquad\quad \cup \\ \qquad\qquad\qquad\qquad\qquad\qquad\qquad\quad S_N \times SU(3). \end{array}$$

This decomposition was carried out for the 3-particle system by Karl and Obryk (15) in the context of the nuclear 3-body problem. It was rederived by Kellman in the process of constructing the rigid-to-nonrigid correlation diagram for three particles (16, 17).

The nonrigid limit selected here, so apt for the nuclear shell model (18), is far from any real molecule because molecules, unlike nuclei, are mostly filled space. The ratio of equilibrium density to hard-core density in a cluster of atoms bound only by van der Waals forces is about o.7 to o.9; the corresponding ratio for nuclei is o.ol to o.1. But this tells us only that molecules must lie toward the rigid side of the diagram; the utility of the selected limit is no less for this.

The "rigid" limit may indeed be a rigid-rotator, small-amplitude vibrator, but it need not. We may, taking the 4-body system as an example, use the model of two stiff diatomics very loosely bound to each other as this limit, or an ammonia-like inverter, just as easily as we can use a square or tetrahedron.

In the "rigid" limit, we assume separability of rotations from vibrations. We also assume the molecule is a spherical top, and that the electronic wavefunction is completely symmetric. We go one step further, introducing an altogether artificial symmetry by supposing for purposes of construction that the normal modes of vibration of the rigid molecule are all degenerate. Artificial though it is, this symmetry is high enough to let us identify all the quantum numbers to connect the rigid and nonrigid limits. At the same time, the artificial symmetry can be broken by well-known means to reduce the symmetry to that of the molecular symmetry (MS) group (6) or the Q-group, if the molecule is nonrigid. (3) Alternatively, the Frobenius reciprocity theorem (19) lets us construct the group of artificially high symmetry from the true MS or Q group. The key steps required here involve the bookkeeping that relates particular representations of the MS group to the irreducible representations of the permutation-inversion group of the N identical particles. The required relations were worked out by Hougen (4), Watson (5) and Bunker (6).

The symmetry of the rigid limit in its idealized form is O(3) x O(3) x SU(3N-6). The O(3) x O(3), isomorphic with O(4), is associated with the $(2J+1)^2$ degeneracy of the j-th level of the spherical top. The SU(3N-6) symmetry is that of the 3N-6-fold degenerate harmonic oscillators representing the normal modes.

The totally symmetric representations of this symmetry group are split by subduction into the representations of the cluster in the region between the two limits, just as were the representations of U(3N-3) for the nonrigid limit. The symmetry group in the intermediate region always contains O(3) due to the rotational invariance of the system.

The correlation diagram is constructed in the conventional way. The totally-symmetric irreducible representations representing the energy states of each limit are laid out sequentially on vertical scales, with the zero-quantum state at bottom. The oscillator spacings are of course equal, and the spacing of the spherical top increase linearly with the quantum number J. The diagram is completed by connecting levels at the two sides, from the bottom of the diagram upward; those levels are connected that belong simultaneously to the same representation of the rotation group O(3), the particle permutation group S_N and the inversion group (E,E^*). The noncrossing rule, that two levels with the same symmetry cannot cross in a one-parameter space, completes the prescription unambiguously.

Decomposition of SU(6) Non-Rigid Limit States under O(3) and S_3

ν π	g_ν	J=0	J=1	J=2	J=3	J=4	J=5	J=6	J=7	J=8
0 +	1	1e 1,0,0								
1 -	6		2o 0,0,1							
2 +	21	3e 1,0,1	1e 0,1,0	3e 1,0,1						
3 -	56		6o 1,1,2	2o 0,0,1	4o 1,1,1					
4 +	126	6e 2,0,2	3e 0,1,1	9e 2,1,3	3e 0,1,1	5e 1,0,2				
5 -	252		12o 2,2,4	6o 1,1,2	12o 2,2,4	4o 1,1,1	6o 1,1,2			
6 +	462	10e 3,1,3	6e 0,2,2	18e 4,2,6	9e 1,2,3	15e 3,2,5	5e 0,1,2	7e 2,1,2		
7 -	792		20o 3,3,7	12o 2,2,4	24o 4,4,8	12o 2,2,4	18o 3,3,6	6o 1,1,2	8o 1,1,3	
8 +	1287	15e 4,1,5	10e 1,3,3	30e 6,4,10	18e 2,4,6	30e 6,4,10	15e 2,3,5	21e 4,3,7	7e 1,2,2	9e 2,1,3
			$\alpha=4$		$\alpha=3$		$\alpha=2$	$\alpha=1$		$\alpha=0$

Figure 1. Decomposition of the group SU(6) for the nonrigid limit of the 3-body problem, including the classification of states according to α, the number of vibrational quanta they contain when carried to their rigid limit.

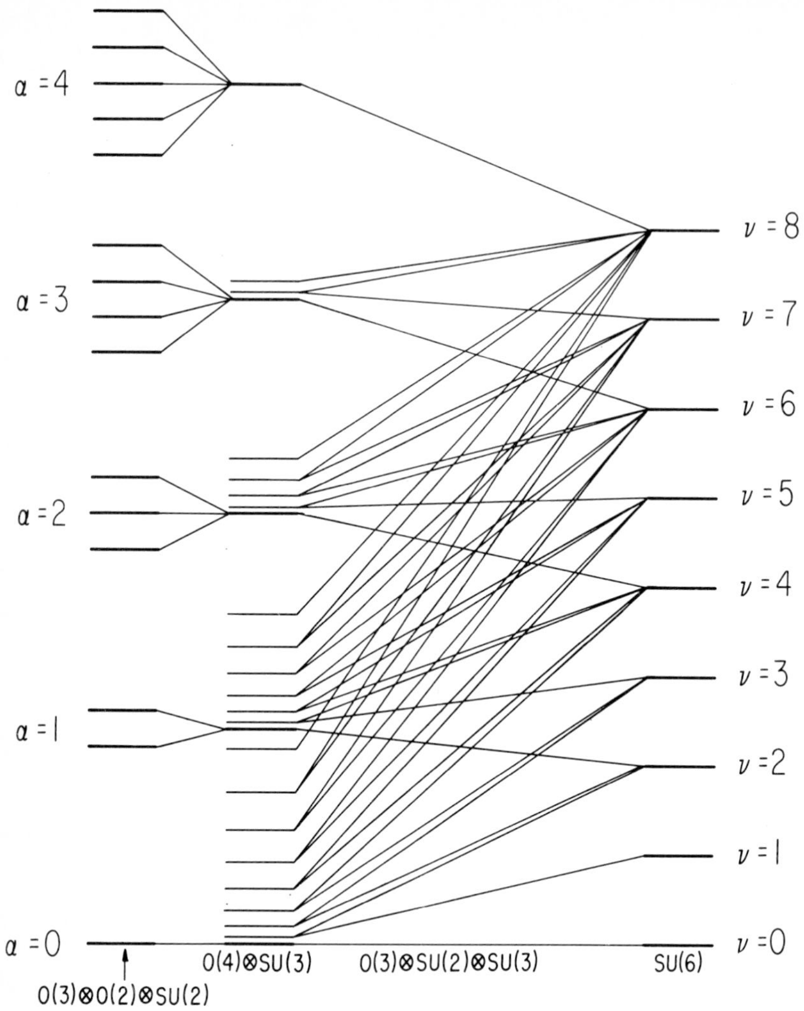

Figure 2. The correlation diagram for the 3-body problem, with the rigid limit of an equilateral triangle at left and the nonrigid limit at right. The ratio of vibration frequencies of the two limits are taken to be slightly greater than 2:1 and the rotational constant for the rigid limit is approximately 0.045 times the vibration frequency for the rigid limit.

Table 1

ν	π	g_ν	J=0	J=1	J=2	J=3	J=4	J=5	J=6
0	+	1	1 0 0 0 0						
1	-	9	0 0 0 0 0	0 0 0 0 1					
2	+	45	1 0 1 0 1	0 0 0 1 0	1 0 1 0 1				
3	-	165	0 1 0 0 0	1 0 1 2 3	0 0 1 1 1	1 0 0 1 2			
4	+	495	3 0 3 1 3	0 1 1 3 2	3 1 4 3 5	0 1 1 2 2	2 0 2 1 2		
5	-	1287	0 1 1 1 0	3 1 4 7 10	2 2 4 6 5	3 1 4 7 9	1 1 2 3 3	1 0 1 2 4	
6	+	3003	6 2 6 4 8	1 3 4 10 7	8 4 13 12 17	3 4 6 11 10	6 3 9 9 12	1 1 3 5 4	3 1 3 2 4

Classifications of the states of the nonrigid limit of the 4-body problem according to their permutation symmetry and total angular momentum quantum number J. The columns at left indicate, in succession, the number of quanta, the parity and the degeneracy of the SU(9) representation. The five rows for each successive column give the number of times each permutation class appears, with the classes in the order corresponding to the Young tableaux $[4]$, $[1^4]$, $[2^2]$, $[2,1^2]$, $[3,1]$, or to A_1, A_2, E, F_1 and F_2.

IV. Correlation Diagrams for Simple Systems

The three-body system with the equilateral triangle as the rigid limit is the simplest nontrivial problem of rigid-to-nonrigid correlation, in the sense of the approach used here. Physical examples for which this analysis is likely to be useful are the very flexible alkali triatomics (2o) such as Na_3, the simple triatomic van der Waals molecules such as Ar_3, and possibly excited states of ozone (21) and similar species. The analysis was worked out previously (16, 17, 22), so we merely summarize the results here.

The decomposition the group SU(6) for the nonrigid limit (the inversion group is in the full group but, for clarity, is excluded from the diagram) is given in Figure 1. The first column gives ν, the number of quanta; the parity π is in the second column, and the degeneracy of the SU(6) representation g_ν is in the third column. Beneath each column head, J, are the number of times each (2J+1)- degenerate representation of SO(3) is included in the νth representation of SU(6); the next three numbers give the composition of this representation in terms of the symmetric, antisymmetric and mixed representations of the symmetric group S_3. Additionally, the "staircase" classification shows the way the irreducible representations connect with α, the number of vibrational quanta, of the rigid-limit SU(3).

The three-body correlation diagram, Figure 2, shows the way the rotational-vibrational ladder of the O(4) x SU(3) rigid limit is tied to the SU(6) nonrigid limit when the vibrational spacing of the rigid limit is roughly twice that of the nonrigid limit and approximately eighty times the rotational constant B of the spherical top. The parameters correspond roughly to what would be expected for Ar_3. At the far left, the splittings are shown for the rigid-limit vibrational levels when the symmetry is reduced to that of the molecular symmetry group.

The four-body problem is richer than the three-body problem partly because one has several reasonable choices for the "rigid" limit. Several were worked out when this system was explored (23). The nonrigid limit is based on a decomposition similar to that of Figure 1, and is given in Table 1. The degeneracies of the ν th representation of SU(3N-3) go up as the binomial coefficient

$$g_\nu = \frac{[(3N-3)+\nu-1]!}{\nu!\,(3N-4)!}$$

so the density of levels goes up rather more rapidly for the 4-body problem than for the three. In the table, the irreducible representations are decomposed according to J and to the cycles or Young tableaux [4], $[1^4]$, $[2^2]$, $[2,1^2]$ and [3,1], with dimensions 1, 1, 2, 3 and 3. (19)

The most obvious rigid limit for four identical particles is the regular tetrahedron, whose molecular symmetry group has irreducible representations A_1, A_2, E, F_1 and F_2 and is isomorphic with the point group T_d and with S_4. The decomposition of the even levels of the idealized

Table 2

α	J=0	J=1	J=2	J=3	J=4	J=5	J=6
	1	0	0	0	1	0	1
	0	0	0	1	0	0	1
0	0	0	1	0	1	1	1
	0	1	0	1	1	2	1
	0	0	1	1	1	1	2
	1	0	2	1	3	2	4
	0	1	1	2	2	3	3
1	1	1	3	3	5	5	7
	0	3	3	6	6	9	9
	1	2	4	5	7	8	10
	3	1	6	4	10	8	13
	0	3	4	7	7	10	11
2	3	4	10	11	17	18	24
	1	10	11	20	22	31	32
	3	7	14	18	24	28	35
	6	4	14	14	24	22	34
	2	8	10	18	20	26	30
3	6	12	26	30	44	50	62
	4	24	32	52	60	80	88
	8	20	36	48	64	76	92

Each column gives the number of levels with A_1, A_2, E, F_1, F_2

Classification of the vibration-rotation states of the idealized rigid tetrahedron. The five rows for each α (number of vibrational quanta) correspond to the classes A_1, A_2, E, F_1 and F_2. Only even levels are shown; the odd levels have an identical table with the subscripts 1 and 2 interchanged on the class labels.

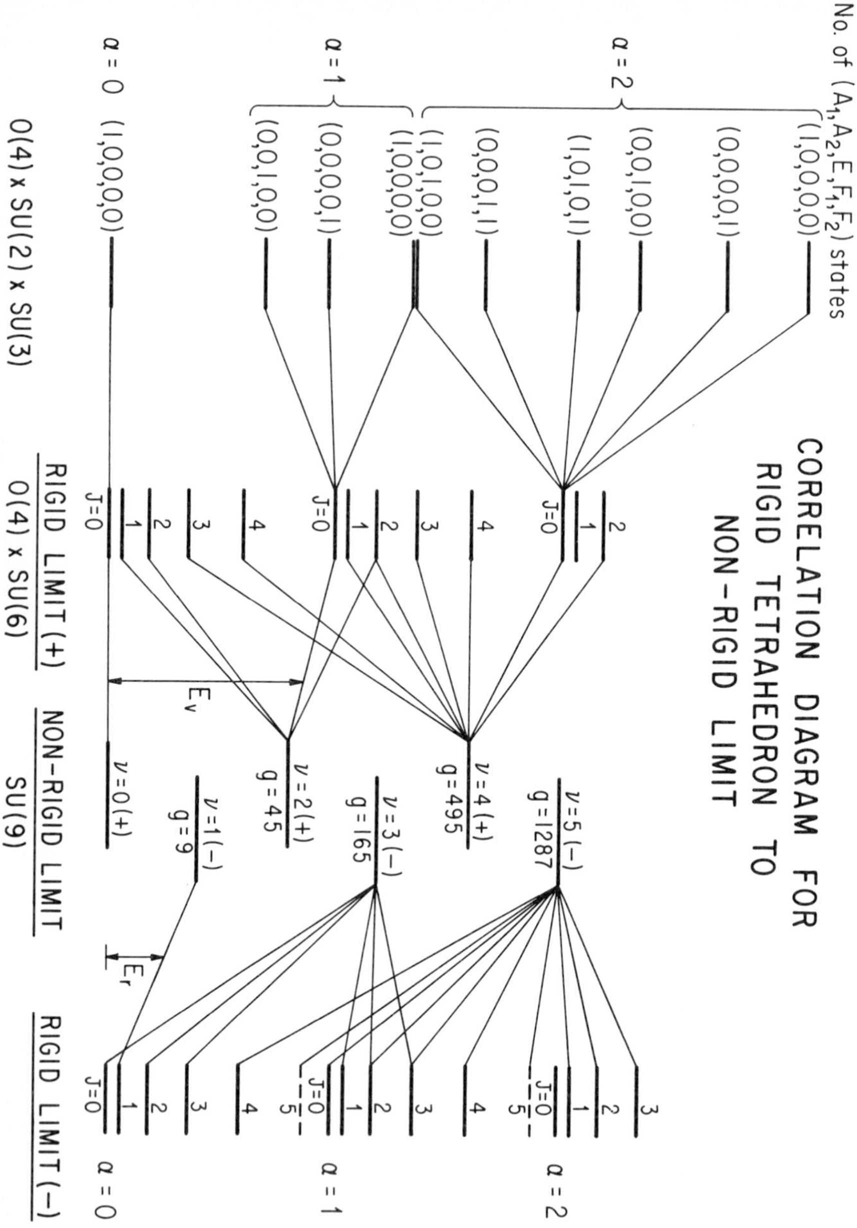

Figure 3. The correlation diagram for the 4-body problem with the tetrahedron as the rigid limit. Odd and even levels of the rigid limit are put on opposite sides of the nonrigid limit, for clarity. At left, the degeneracy of the artificial SU(6) symmetry of the rigid limit is lifted.

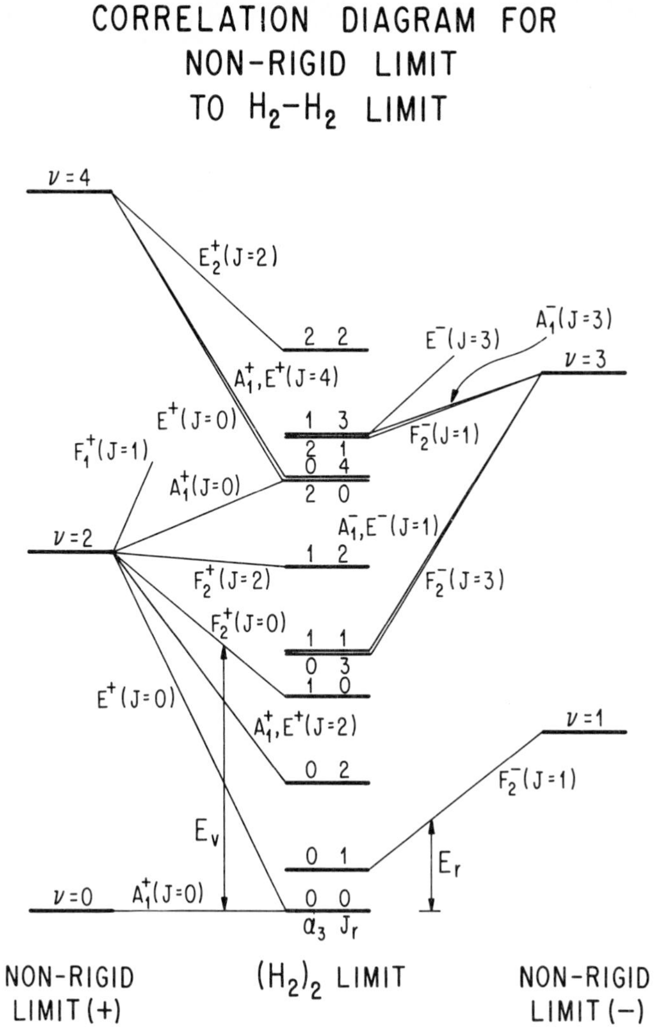

Figure 4. The correlation diagram for the 4-body problem with the diatomic dimer, e.g. $(H_2)_2$, as the "rigid" limit. The even and odd levels of the nonrigid limit are shown on opposite sides of the "rigid" limit, for clarity. The figure is taken from Ref. 23.

rigid tetrahedron according to J and the molecular symmetry group representations (in the order just given) is presented in Table 2. The odd levels are expressed by the same table with the subscripts 1 and 2 interchanged. The completed correlation diagram for the tetrahedral limit is in Figure 3. The odd and even levels of the rigid limit are shown on opposite sides of the nonrigid limit. At the far left are the levels of the irrotational states split from the idealized SU(6) limit to the MS group.

A third example is given in Figure 4. Here we see the correlation diagram of a system whose "rigid" limit is two H_2 molecules. This time the odd and even levels of the nonrigid limit are split to opposite sides of the $(H_2)_2$ limit. The quantum numbers of the H_2-H_2 vibration and overall (total) rotational are given for the $(H_2)_2$ limit.

The question arises naturally as to whether one can give any quantitative meaning to the abscissa of these diagrams. This is the same as asking whether it is possible to find a single parameter to represent the amount of nonrigidity in a molecule. Yamada and Winnewisser (24) introduced a parameter to measure deviations from nonlinearity--a sort of nonrigidity--for polyatomic chain molecules. Their parameter is the ratio of two energies, the energy of excitation of the first excited state with J = 1 to the corresponding energy for the first excited state with J = 0.

We can extend this definition, with the help of our correlation diagrams, to more general polyatomic molecules. Referring to the <u>rigid</u> limit, we call E_r^0 the energy of excitation to the state that correlates with the first rotationally-excited, vibrationally unexcited level. Correspondingly, we call E_v^0 the energy of excitation to the state correlating with the first vibrationally-excited, rotationless state in the rigid limit. The limit of E_r^0 in the nonrigid limit is $h\omega$, and the limit of E_v^0 in the nonrigid limit is $2\hbar\omega$, where ω is the characteristic frequency of the pairwise harmonic interaction of that limit. Now we construct the ratio

$$\gamma = 2E_r^0/E_v^0.$$

Table 3.

Values of the nonrigidity parameter γ for representative triatomic and tetratomic molecules.

Molecule	γ	Remarks
O_3	0.0061	E_v^o = ave of normal modes
Ar_3	0.013	$E_v^o \sim 20^0K$
P_4	0.00096	E_v^o = ave of all fundamentals
Ar_4	0.00772	$E_v^o = 15\ cm^{-1}$, $E_r^o = 0.058\ cm^{-1}$
NH_3	0.01328	E_v^0 = ave of all fundamentals
NH_3	0.0336	E_v^o = inversion fundamental only
CH_4	0.01274	E_v^o = ave of all fundamentals
$(H_2)_2$	0.40	$E_v^o = 10\ cm^{-1}$; $E_r^0 = 2\ cm^{-1}$

If we consider the conceptual limit of the rigid case one for which the vibrational frequency grows without bound, then we have $0<\gamma<1$, with 0 and 1 representing the values of γ for rigid and nonrigid limits, respectively. Naturally we should not expect too much of one parameter; γ carries the burden of representing as much as we can put into one number about the floppiness of a molecule. The value of γ is model-dependent because the states that comprise any given level everywhere in the diagrams (except the nonrigid limits) depend on the choice of "rigid" limit. To calculate γ, we choose a model, find what levels of the real molecule correlate, in the intermediate region, into the state with energy E_v^0 and similarly for the state with energy E_r^0. The statistically weighted averages of those impirical excitation energies give the values of E_v^0 and E_r^0, from which we determine γ. Values of γ for several molecules (22, 23) are given in Table 3.

The value of γ does not betray the mechanisms of rearrangement nor the "tunneling rate." In our picture, no tunneling process appears because we invoke no mechanistic potential surface to carry out the phenomenology. Note however that one could quite well construct levels and evaluate a predicted γ if one chose to start with a surface based on electronic energy levels and the Born-Oppenheimer approximation. The relative rigidity of molecules is clear from the small values of γ for almost all the molecules listed in Table 3. That is, they almost all fall near the rigid limit in their correlation diagrams.

V. Representation of Spatial Correlation

One of the tangential questions that grew from our inquiry into correlation the tendency of nonrigid clusters to take on polyhedral geometry. How could one detect and quantify such a tendency? whatever approach one uses, one must examine interparticle angles, not merely interparticle distances, in order to come to any answer to this question. The simplest nonrigid structures in which one might look for such effects are the two-electron, helium-like atoms. Studies of the doubly-excited states of helium (25-28) suggested that these states might indeed show such tendencies. We found, in our explorations (29), that it is not a formidable task to construct $\rho(\theta_{12})$, the distribution function for the angle θ_{12} defined by the position vectors r_1 and r_2 of electrons 1 and

Fig. 5 a

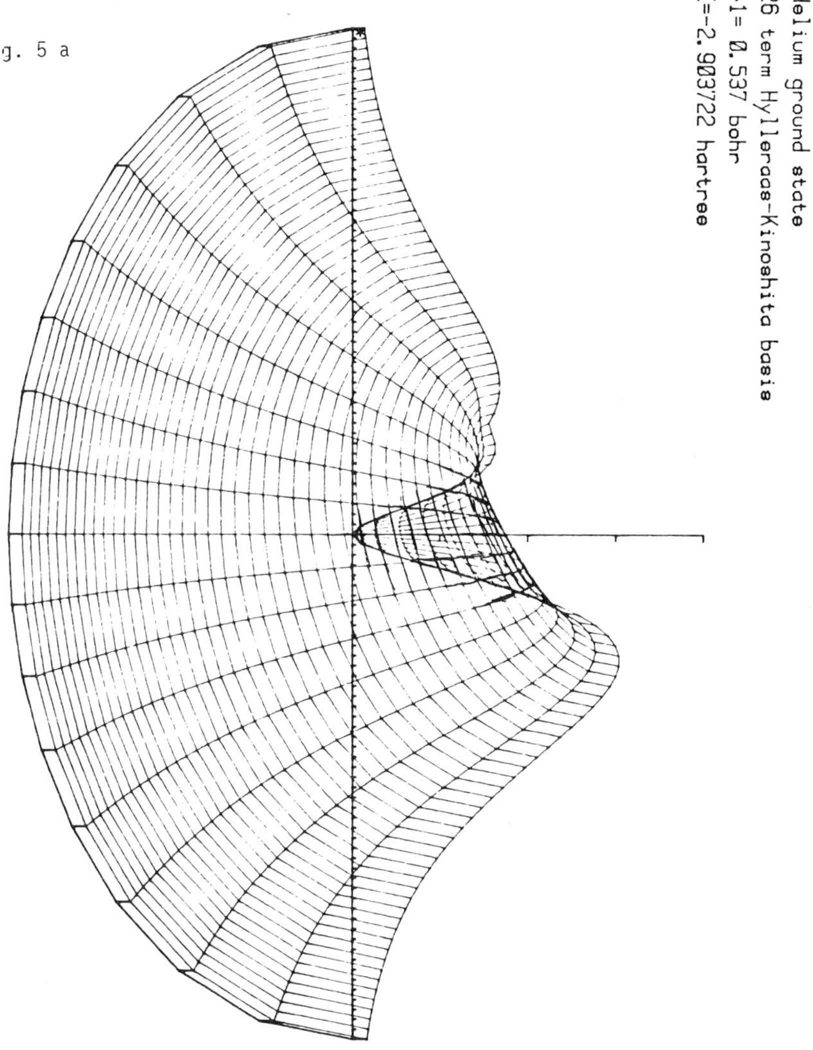

Figure 5. Conditional probability densities $d(R_2,\theta_{12}|r_1^o)$ for the ground states of He and H^-. The values of r_1^o, o.537 and 1.178 bohr, are the most probable values for r_1 in the two cases. The first and third figures were computes with 26-term Hylleraas-Kinoshita wavefunctions giving energy values within $1o^{-7}$ of the values obtained by Pekeris (Phys. Rev. 126, 147o (1962)). The second figure is based on the many-configuration function of Holøien. (31) All three figures are taken from Ref. 3o.

Fig. 5 b

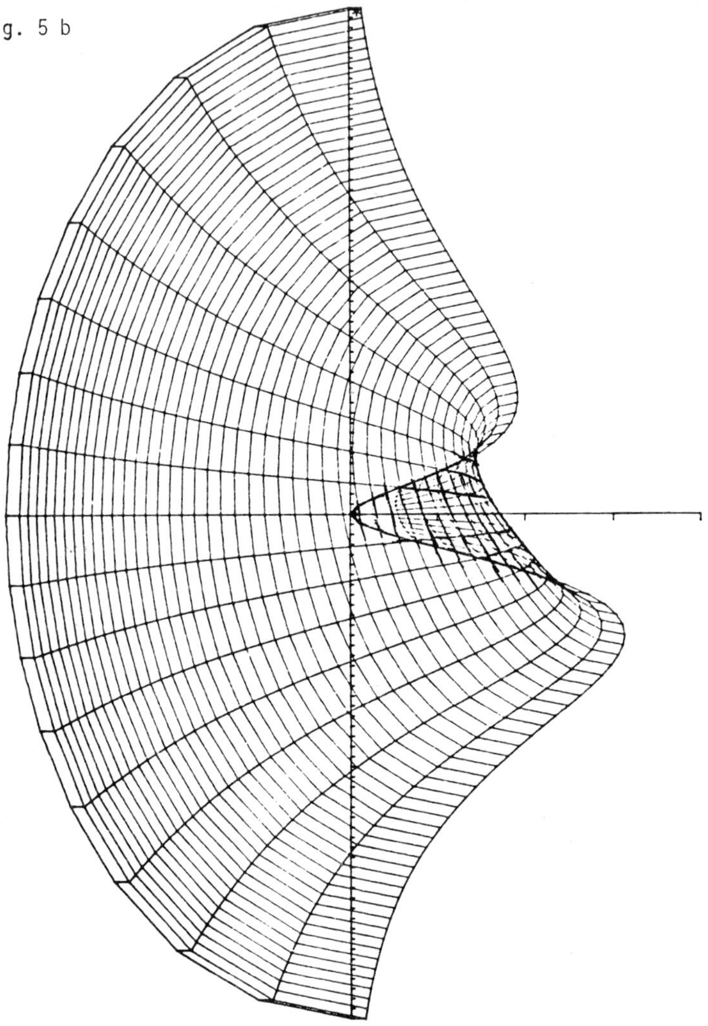

20 Configuration (CI) basis
E=-2.90123145 hartree

Fig. 5 c

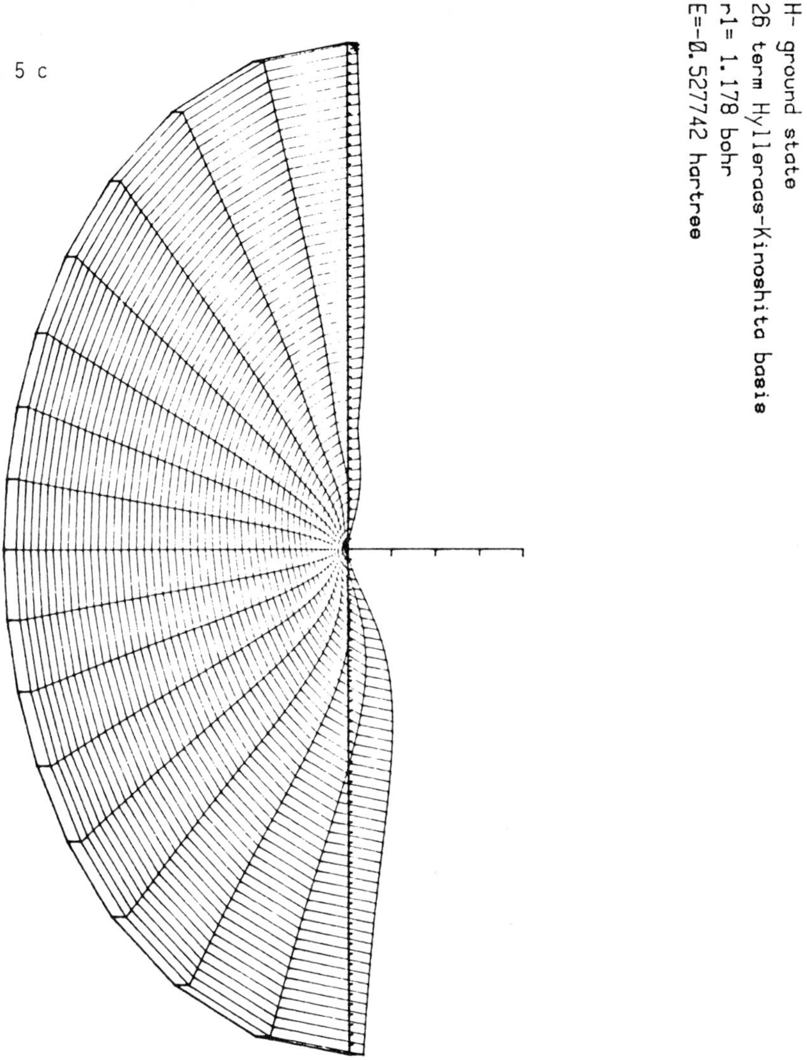
H- ground state
26 term Hylleraas-Kinoshita basis
r1= 1.178 bohr
E=-0.527742 hartree

2 from the nucleus. However we found that one could even more easily generate the more powerful function $d(r_2,\theta_{12}|r_1^0)$, the conditional probability for finding electron 2 at distance r_2, and at the angle θ_{12} from r_1, when electron 1 is at distance r_1^0. By constructing graphs of $d(r_1,\theta_{12}|r_1^0)$ for various values of r_1^0, we could distinguish the conditions--e.g. the ranges of dr_1^0--for which there is a tendency towards polyhedral disposition of the electrons; the breadth of the distributions give clear visual indications of how strong the tendency is.

Some aspects of the information obtained by such analyses is illustrated by the projected 3-dimensional graphs of ground states of He and H^- of Figure 5 (3o). The conditional probability density is the vertical coordinate; the angular coordinate θ_{12} is measured outward from that point. Electron 1 is fixed at its most probable distance, o.537 bohr for He and 1.178 bohr for H^-, along the axis $\theta_{12} = 0$. Figure 5 shows how much more "swollen" is the negative hydride ion than the isoelectronic He atom; it also shows, in the first two drawings, how a very good configuration interaction function (31) reproduces the long-range effects of correlation, here a tendency toward linearity, but entirely fails to represent the very short-range correlation effects. Only with the interparticle distance incorporated in the wavefunction do we see the "kink" due to the cusp at $r_{12} = 0$; this appears in the first and third but not in the second drawing of Figure 4. The kinks are precisely the instantaneous holes around electron 2 due to the Coulomb repulsion of the two electrons. (This kind of Coulomb hole, which corresponds to the intuitive concept of something in the vicinity of the instantaneous position $r_{12} = 0$, should not be confused with the definition used by Coulson and Neilson (32), as the difference between the distributions $\rho_{HF}(r_{12})$ and $\rho_{corr}(r_{12})$, based on Hartree-Fock and correlated wavefunctions, respectively.)

The analyses show that there is some tendency, especially in excited states, (29. 3o, 33) for recognizable geometries to be established in two-electron systems. The next stage, the extension of such analyses to clusters of three identical particles with short-range rather than Coulomb interactions, will bring to the study of nonrigid structures a tool that will let us visualize what we could previously only intuit from such cues as the correlation diagrams.

VI. Symmetry and the Melting and Nucleation of Clusters

The construction of the correlation diagrams such as those of Figures 2-4 gives us a means to evaluate approximate densities of states for rigid and nonrigid clusters of N identical particles. (If we accept the straight-line connections, we can construct densities of states for all values of γ between the limits.) From the densities of states, we can determine partition functions q_{rig} and q_{nonr} for the rigid and nonrigid clusters:

$$q_{rig} = \frac{\pi^{1/2}}{\sigma}\left(\frac{T}{B_e}\right)^{3/2}\left(\frac{e^{-V/2T}}{1-e^{-V/T}}\right)^{3N-6} e^{D_e/T}$$

and

$$q_{nonr} = \frac{1}{N!}\left(\frac{e^{-W/2T}}{1-e^{-W/T}}\right)^{3N-3} e^{D_e'/T}$$

Here, σ is the symmetry number of the rigid N-body molecule, B_e is its rotational constant (in units of ^oK), D_e is its binding energy, and V is its (single) vibrational frequency (also in ^oK); W is the corresponding vibrational frequency for the nonrigid cluster, and D_e' is its binding energy. The factor of 1/N! avoids overcounting the vibrational phase space. If the mean particle displacement on the time scale of the observation is small enough to obviate the establishment of any particle equivalent, then this factor should be dropped. (If the experimental time peak is long enough for establishment of some but not complete particle equivalence, then it might be possible in specific cases to assign an "equivalence number $\eta = \prod_{j=1}^{r} N_j!$ such that η^{-1} represents the removal of overcounting due to establishment of equivalences among groups 1,...,j,...,r of identical particles.) The Helmholtz free energy is $A = -kT\ln q$.

From the Helmholtz free energy, we now compute two properties, the "melting temperature T_m" at which the rigid and nonrigid cluster of N particles have equal Helmholtz free energies, as a function of N, and the critical cluster size N_{cr}, the lowest N for which the free energy of cluster-plus-free-particle is greater than that of the cluster of N+1 particles.

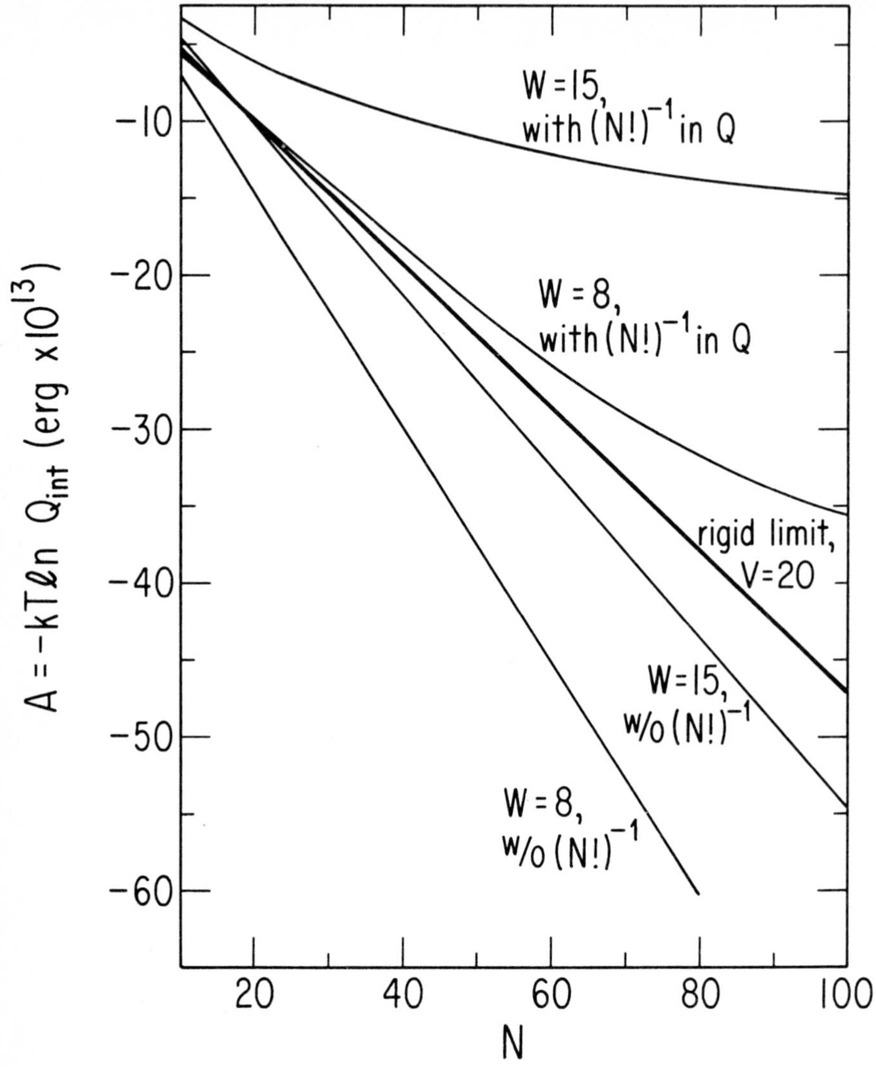

Figure 6. The behavior of the Helmholtz free energy $A = -kT\ell nQ$ as a function of the number of particles in the cluster for the rigid limit (heavy line) and the nonrigid limit for two normal mode frequencies, with and without the $(N!)^{-1}$ factor that is included when the N particles are equivalent on the time scale of the observation.

Fig. 7 a

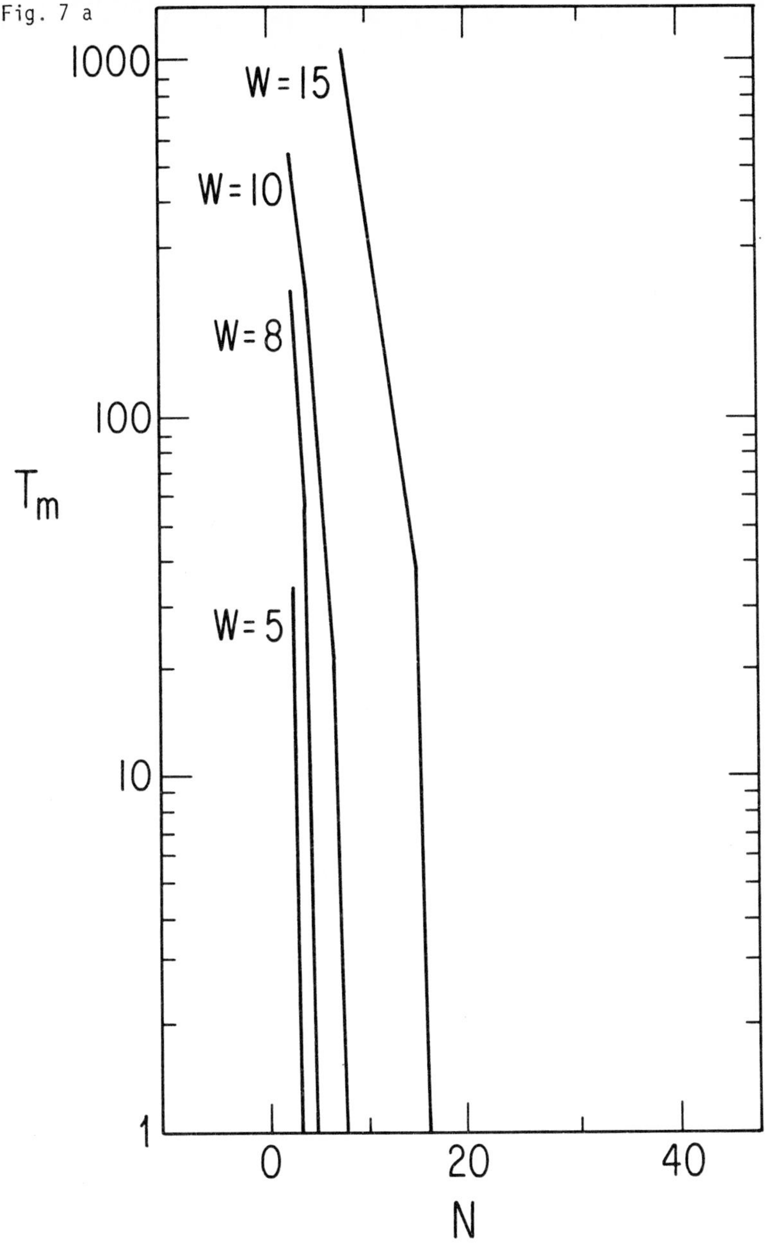

Figure 7. Melting temperatures vs. N for various values of the frequency W, a) without $(N!)^{-1}$ in the partition function, and b) with $(N!)^{-1}$ in the partition function. Both graphs were drawn for symmetry numbers of 1 for the rigid limit, and no heat of melting; with a heat of melting of 142^{0} K, V of $2o^{0}$ K and no factor of $(N!)^{-1}$, the curves of c) are obtained. These have minima at the points marked with arrows. The crosses mark melting points computed from molecular dynamics (Ref. 38).

Fig. 7 b

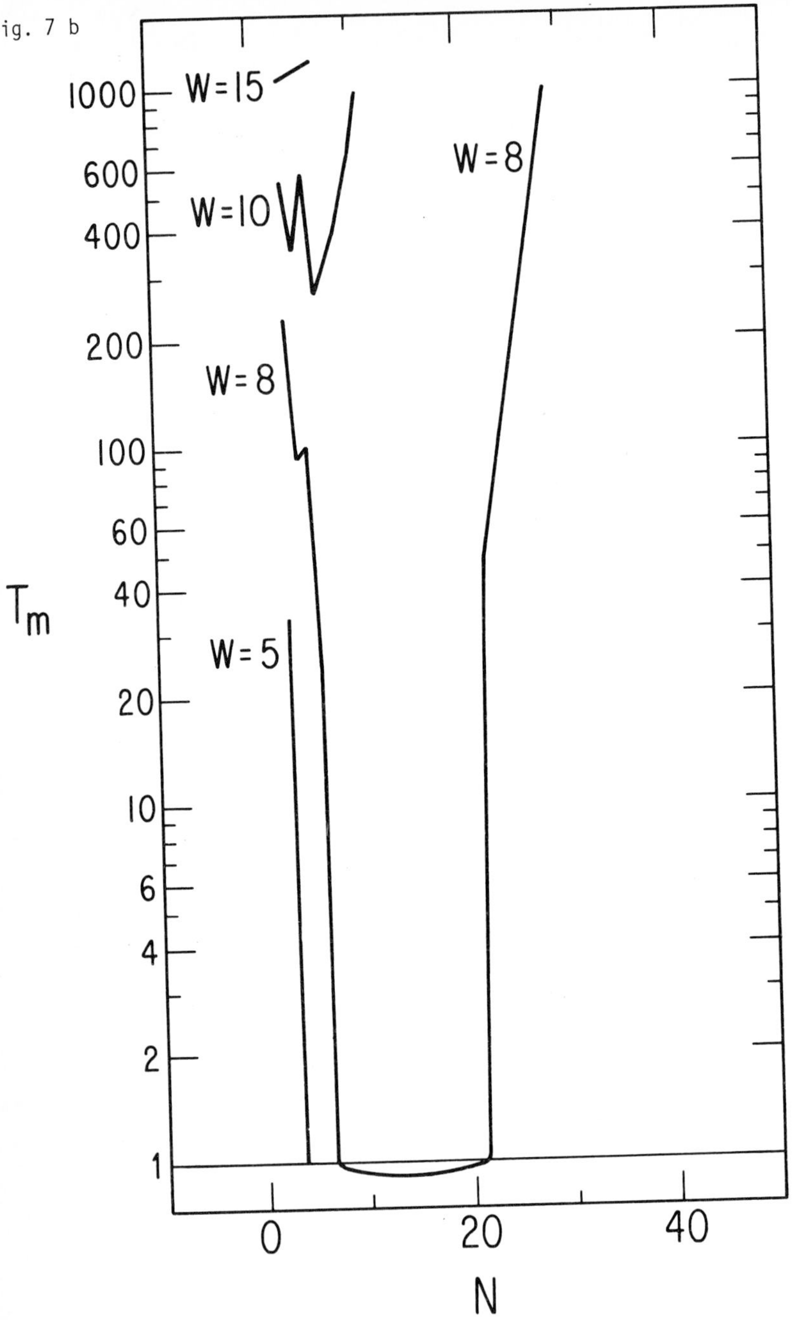

Fig. 7 c

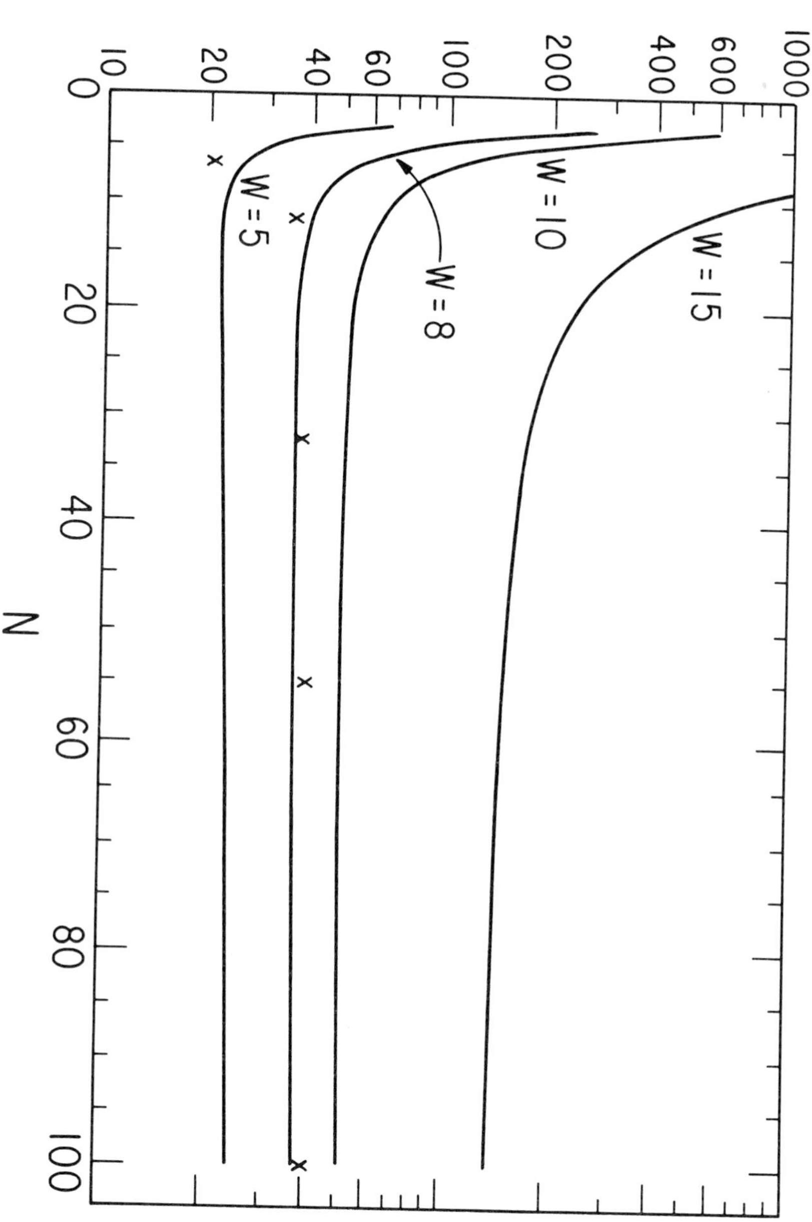

The "melting temperature" of small clusters has become a matter of increased interest since the computer simulations of McGinty, (35, 36) Lee, Barker and Abraham, (37) Briant and Burton (38) and Etters and Kaelberer. (39) Both Monte Carlo and molecular dynamics simulations indicate that small clusters exhibit both liquid-like and solid-like behavior. The simulations show, furthermore, rather sharp transitions, between rigid and nonrigid forms. The results are strongly suggestive that for all but the smallest clusters, the range of detectable coexistence of the two "phases" would be so narrow that we would recognize the passage from one to the other as a phase transition.

Taking an equilibrium pair distance of 3.4A for our model, based on a 6-12 potential for argon, (4o) we obtain a rotational constant (in Kelvin)

$$B_e = 0.412N^{-5/3}$$

for a spherical cluster. The vibrational spacing v of the rigid cluster is assumed 20°K. Figure 6 shows the behavior of $A_{rig}(N)$ and $A_{nonr}(N)$ as functions of N, for these parameters with the frequency W taken to be 8 and 15°K. In Figure 7a is a plot of T_m, the temperature at which $A_{rig}=A_{nonr}$, for several values of W, and with $\sigma=1$ and no factor of $(N!)^{-1}$, corresponding to distinguishable particles. Figure 7b shows the behavior of $T_m(N)$ at the other extreme, with σ given its largest value for a symmetric polyhedron and 1/N! included as a factor in q_{nonr}. Both of these were computed with no dissociation energy for the liquid$\leftrightarrow$solid transition, that is, with $\Delta D=D_e-D_e'=0$. The conclusion is that the value of T_m is sensitive to the ratio V/W, but that with reasonable estimates for these frequencies one can match the behavior of $T_m(N)$ for computer simulations reasonably well, over a rather wide range of N. The statistical-mechanical model contains several parameters, but all of them, with the possible exception of ΔD, are available by procedures that do not involve the melting process.

Acknowlegements. This work represents the results of the doctoral research of M.E. Kellman, F. Amar and P. Rehmus. Refs. 1o, 17, 22, 23, 29, 3o, 33 and 34 describe this work in more detail. The research was supported by the National Science Foundation of the United States.

REFERENCES

1. R.S. Berry, Revs. Mod. Phys. 32, 447 (196o).
2. R.G. Woolley, J. Am. Chem. Soc. 1oo, 1o73 (1978).
3. H.C. Longuet-Higgins, Molec. Phys. 6, 445 (1963)
4. J.K.G. Watson, Can. J. Phys. 43, 1996 (1965).
5. J.T. Hougen, J. Chem. Phys. 37, 1433 (1962); ibid. 39, 358 (1963); ibid 55, 1122 (1971).
6. P.R. Bunker, in Vibrational Spectra and Structure, edited by J.R. Durig (Marcel Dekker, New York, 1976); Molecular Symmetry and Spectroscopy (Academic Press, New York, 1978).
7. B.J. Dalton, J. Chem. Phys. 54, 4745 (1971).
8. E.L. Muetterties, Accts. Chem. Res. 3, 266 (197o).
9. K. Mislow, Accts. Chem. Res. 3, 321 (197o).
1o. I. Ugi, Angew. Chem. Int'l. Ed. 1o, 637 (1971).
11. P. Ehrenfest, Proc. Acad. Sci. Amsterdam 16, 591 (1914); Naturwiss. 27, 543 (1928); H.A. Kramers, Quantum Mechanics (North-Holland, Amsterdam, 1957), p. 215.
12. S. Gartenhaus and C. Schwartz, Phys. Rev. 1o8, 482 (1957).
13. G.A. Baker, Phys. Rev. 1o3, 1119 (1956).
14. P. Kramer and M. Moshinsky, Nucl. Phys. 82, 241 (1966).
15. G. Karl and E. Obryk, Nucl. Phys. B8, 6o9 (1968).
16. M.E. Kellman, Doctoral Dissertation, University of Chicago, 1977.
17. M.E. Kellman and R.S. Berry, Chem. Phys. Lett. 42, 327 (1976).
18. M.A. Preston, Physics of the Nucleus (Addison-Wesley, Reading, Mass., 1962).
19. See, for example, S.L. Altmann, Induced Representations in Molecules and Crystals (Academic Press, London, 1977), M.J. Petrashen and E.D. Trifonov, Applications of Group Theory in Quantum Mechanics (M.I.T. Press, Cambridge, Mass., 1969), or M. Hamermesh, Group Theory (Addison-Wesley, Reading, Mass., 1962).
2o. A. Hermann, S. Leutwyler and E. Schumacher, Helv. Chim. Act. 61, 453 (1978).
21. S.D. Peyerimhoff and R.J. Buenker, J. Chem. Phys. 47, 1953 (1967); N. Swanson and R.J. Celotta, Phys. Rev. Lett. 35, 783 (1975).
22. M.E. Kellman, F. Amar and R.S. Berry, "Correlation Diagrams for Rigid and Nonrigid 3-body Systems," (to be published).

23. F. Amar, M.E. Kellman and R.S. Berry, "Correlation Diagrams for Rigid and Nonrigid 4-body Systems," J. Chem. Phys. 70 , 1973 (1979).
24. K. Yamada and M. Winnewisser, Z. Naturforsch. Teil A, 31, 134 (1976).
25. C. Wulfman and J. Kumei, Chem. Phys. Lett. 23, 367 (1973).
26. C. Wulfman, Chem. Phys. Lett. 23, 37o (1973).
27. O. Sinanoglu and D.R. Herrick. J. Chem. Phys. 62, 886 (1973).
28. D.R. Herrick and O. Sinanoglu, Phys. Rev. A 11, 97 (1975).
29. P. Rehmus, M.E. Kellman and R.S. Berry, Chem. Phys. 31, 239 (1978).
3o. P. Rehmus, C.C.J. Roothaan and R.S. Berry, Chem. Phys. Lett. 58, 321 (1978).
31. E. Holøien, Proc. Phys. Soc. 71, 141, 357 (1958).
32. C.A. Coulson and A.H. Neilson, Proc. Phys. Soc. 78, 831 (1961).
33. P. Rehmus and R.S. Berry, Chem. Phys. 38 , 257 (1979).
34. F. Amar, M.E. Kellman and R.S. Berry (in preparation).
35. D.J. McGinty, J. Chem. Phys. 55, 58o (1971).
36. D.J. McGinty, J. Chem. Phys. 58, 4733 (1973).
37. J.K. Lee, J.A. Barker and F.F. Abraham, J. Chem. Phys. 58, 3166 (1973).
38. C.L. Briant and J.J. Burton, J. Chem. Phys. 63, 2o45 (1975).
39. R.D. Etters and J. Kaelberer, J. Chem. Phys. 66, 3233, 5112 (1977).
4o. G.E. Ewing, Canad. J. Phys. 54, 437 (1976).

REPRESENTATIONS OF THE SYMMETRIC GROUP AS SPECIAL CASES OF THE BOSON POLYNOMIALS IN U(n)

L.C. Biedenharn *
Physics Department, Duke University
Durham, North Carolina 27706, USA

J.D. Louck **
Group T-7, Theoretical Division
Los Alamos Scientific Laboratory
Los Alamos, New Mexico 87545, USA

Abstract

The set of all real, orthogonal irreps of S_n are realized explicitly and non-recursively by specializing the boson polynomials carrying irreps of the unitary group. This realization makes use of a 'calculus of patterns', which is discussed.

Introduction

The purpose of the present paper is to show how some recent investigations in the unitary group -- motivated by applications to quantum physics -- can be specialized to yield interesting -- and we hope, useful -- results for the symmetric group. Our main result is to obtain an explicit, non-recursive, set of real, orthogonal irreps for S_n, whose realization by means of the pattern calculus (as explained below) is (we believe) new.

Let us indicate, very briefly, why the unitary group figures so prominently in quantum physics. The *state*, ψ, of a quantal system is a *ray* in Hilbert space of unit length; *observables* are self-adjoint operators, O, mapping the Hilbert space into itself. A *symmetry* is a mapping of states into states, and operators into operators such that the probability $|\langle\phi|\sigma|\psi\rangle|$ is preserved. The fundamental theorem (Wigner-Artin) -- essentially the fundamental theorem of projective geometry -- now states:

---

* Paper presented by L.C. Biedenharn. Work supported, in part, by the National Science Foundation.

**Work performed under the auspices of the USERDA.

any symmetry can be implemented by a semi-linear unitary transformation. It follows that the unitary group is of basic interest in quantum physics.

Let us remark also that the study of the symmetric group by broadening the investigation to the unitary group is itself a familiar technique; it was used extensively by Weyl, and is one of the principal themes in G. de B. Robinson's monograph on S_n.

It is necessary to explain now precisely what is meant by a "boson", and by a "boson operator". These terms are physicist's jargon for concepts known to mathematicians as the Weyl algebra, (or as it is also called the generators of the Heisenberg group). The boson, a, and its conjugate, $\bar{a}$, are elements of an algebra (Weyl algebra) satisfying the commutation rule: $(\bar{a},a)=1$, where 1 is the unit operator. More generally, we consider n bosons: a_i, $i=1,2,\ldots n$ and their conjugates: $\bar{a}_i$, $i=1,\ldots n$ obeying the rules:

$$(a_i,a_j) = (\bar{a}_i,\bar{a}_j) = 0;\ (\bar{a}_i,a_j) = \delta_{ij}\ .$$

(The name "boson" contrasts with the physicist's term "fermion", which replaces commutation in the rules above by anti-commutation.)

Boson polynomials are simply polynomials (over $\mathbb{C}$) with the bosons $\{a_i\}$ as indeterminates. There is a natural scalar product associated to the boson polynomials by the commutation rule, if we define the abstract vector $|0\rangle$ to be annihilated by all conjugate bosons: $\bar{a}_i|0\rangle \equiv 0$. Then to the boson monomial $(a_i)^k$, we associate the Hilbert space vector: $(a_i)^k|0\rangle \equiv |\psi\rangle$, and the scalar product: $\langle\psi|\psi\rangle = \langle 0|(\bar{a}_i)^k(a_i)^k|0\rangle = k!$.

We remark that the technique of boson operator construction can be phrased in the language of the umbral calculus.

II. *Young tableaux, Weyl tableaux, and Gel'fand patterns*

One of the first problems that one confronts in discussing the irreducible representations of the unitary group U(n) is that of devising a comprehensible notation. This problem was solved in an elegant way by Gel'fand and Zetlin (1) by utilizing the Weyl branching law (2) for U(n). In order to explain this notation in familiar terms, it is convenient to appeal to the concept of standard Young tableaux of the symmetric group S_n since the relationship of these tableaux (to the irreducible representations of S_n) is well known to the participants of this conference.

The first concept required is that of a *Young frame*: a Young frame $Y_{(\lambda)}$ of *shape* $(\lambda)=(\lambda_1\lambda_2\cdots\lambda_n)$, where the λ_i are non-negative integers satisfying $\lambda_1 > \lambda_2 > \ldots > \lambda_n$, is a diagram consisting of λ_1 boxes (nodes) in row 1, λ_2 boxes in row 2,..., λ_n boxes in row n, arranged as illustrated in Fig. 1.

Fig. 1

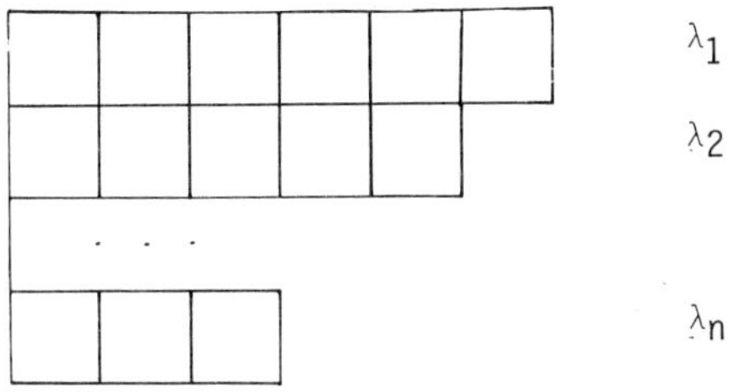

(1)

A *Weyl tableau* is a Young frame in which the boxes have been "filled in" with integers selected from 1, 2,...,n. A Weyl tableau is *standard* if the sequence of integers appearing in each row of $Y_{(\lambda)}$ is *nondecreasing* as read from left to right and the sequence of integers appearing in each column is *strictly increasing* as read from top to bottom. The *weight* or *content* (W) of a Weyl tableau $Y_{(\lambda)}$ is defined to be the row vector $(W)=(w_1,w_2,\ldots,w_n)$, where w_k equals the number of times integer k appears in the pattern. If $\lambda_1+\lambda_2+\ldots+\lambda_n=N$, then also $w_1+w_2+\ldots+w_n=N$. We shall call (λ) a *partition* of N into n parts, or more often, a partition when N is unspecified. We generally count the 0's in determining the parts of a partition. For example, the partitions of 4 into 3 parts are (4 0 0), (3 1 0), and (2 2 0). When the number of parts is understood, one frequently omits the zeroes (writing (4), (3 1), and (2 2) in the examples).

Example. The standard Weyl patterns corresponding to the Young frame  are:

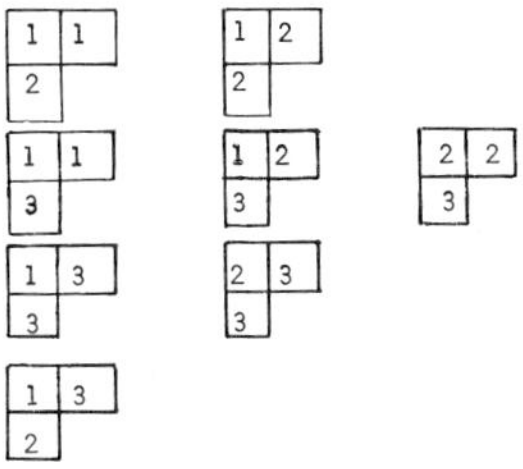

(2)

Young's (3) interest was in invariant theory, utilizing the symmetric group, and he considered frames with n nodes filled in with integers 1 to m. To our knowledge Weyl (4) was the first to use Young frames filled in with repeated integers. We therefore refer to these latter tableaux as *Weyl tableaux*, reserving the term *Young tableaux* for the more restricted case.

Gel'fand patterns. An elegant geometrical notation for codifying the constraints imposed on the entries of a Young pattern is provided by a Gel'fand pattern which we now define.

A Gel'fand pattern is a triangular array of n rows of integers, there being one entry in the first row, two entries in the second row, ..., and n entries in the nth row. The entries in each row 2, 3,..., n-1, are arranged so as to fall between the entries in the row above and below, as illustrated below:

$$\begin{pmatrix} [m] \\ (m) \end{pmatrix} = \begin{pmatrix} m_{1n} & & m_{2n} & & \cdots & & m_{nn} \\ \cdot & \cdot & & & & \cdot & \\ & \cdot & \cdot & & & \cdot & \\ & & \cdot & \cdot & & \cdot & \\ & m_{13} & & m_{23} & & m_{33} & \\ & & m_{12} & & m_{22} & & \\ & & & m_{11} & & & \end{pmatrix} \qquad (3)$$

The integral entries m_{ij}, $i \leqslant j = 1,2,\ldots,n$ in this array are required to satisfy the following rules:

$$(i) \quad m_{1n} \geqslant m_{2n} \geqslant \cdots \geqslant m_{nn}; \qquad (4)$$

(*ii*) for each specified partition $(m_{1n} \ldots m_{nn})$, the entries in the remaining rows j=n-1, n-2,...,1 may be any integers which satisfy the "betweenness conditions"

$$m_{1j+1} \geqslant m_{1j} \geqslant m_{2j+1} \geqslant m_{2j} \geqslant m_{3j+1} \geqslant m_{3j} \geqslant \cdots \geqslant m_{jj+1} \geqslant m_{jj} \geqslant m_{j+1j+1}. \qquad (5)$$

These betweenness conditions are, in fact, just the Weyl branching law for the chain of unitary subgroups given by:

$$U(n) \supset U(n-1) \supset \ldots \supset U(2) \supset U(1) \quad . \qquad (6)$$

Example. For n = 3, and $(m_{13}m_{23}m_{33})$ = (2 1 0), there are eight Gel'fand patterns as displayed below:

$$\begin{pmatrix} 2\ 1\ 0 \\ 2\ 1 \\ 2 \end{pmatrix} \quad \begin{pmatrix} 2\ 1\ 0 \\ 2\ 1 \\ 1 \end{pmatrix}$$

$$\begin{pmatrix} 2\ 1\ 0 \\ 2\ 0 \\ 2 \end{pmatrix} \quad \begin{pmatrix} 2\ 1\ 0 \\ 2\ 0 \\ 1 \end{pmatrix} \quad \begin{pmatrix} 2\ 1\ 0 \\ 2\ 0 \\ 0 \end{pmatrix}$$

$$\begin{pmatrix} 2\ 1\ 0 \\ 1\ 0 \\ 1 \end{pmatrix} \quad \begin{pmatrix} 2\ 1\ 0 \\ 1\ 0 \\ 0 \end{pmatrix} \qquad (7)$$

$$\begin{pmatrix} 2\ 1\ 0 \\ 1\ 1 \\ 1 \end{pmatrix}$$

Mapping between Gel'fand patterns and Standard Weyl tableaux.

There is a one-to-one correspondence between the set of Gel'fand patterns (m) having nth row $(m_{1n}m_{2n}\cdots m_{nn})$ (with $m_{nn} > 0$) and the set of standard Weyl tableaux of this shape.

The mapping between Gel'fand patterns and standard Weyl tableaux is described as follows: The shape of the frame is $(m_{1n}m_{2n}\cdots m_{nn})$, and the rows of the frame are filled in according to the followind rules (read along the diagonals of the Gel'fand pattern):

row 1: m_{11} 1's, $m_{12}-m_{11}$ 2's, $m_{13}-m_{23}$ 3's $\ldots$, $m_{1n}-m_{1n-1}$ n's

row 2: m_{22} 2's, $m_{23}-m_{22}$ 3's, $\ldots$, $m_{2n}-m_{2n-1}$ n's

.
.
.

row j: $\quad m_{jj}$ j's, $\quad m_{jj+1}-m_{jj}$ (j-1)'s,...,$m_{jn}-m_{jn-1}$ n's $\qquad$ (8)

.
.
.

row n: $\qquad m_{nn}$ n's .

Using the rule (8), we see that the set of Gel'fand patterns (7) is mapped to the set of Weyl tableaux (2). Conversely, from each standard Weyl tableau (2), we construct in an obvious way the Gel'fand pattern in the set (7).

The *weight* or *content* of a Gel'fand pattern (m), is the row vector $(W) = (w_1 w_2 \ldots w_n)$, where w_j is defined to be the sum of the entries in row j of (m) minus the sum of the entries in row j-1 ($w_1 = m_{11}$):

$$w_j = \sum_{i=1}^{j} m_{ij} = \sum_{i=1}^{j-1} m_{ij-1} \quad . \tag{9}$$

Clearly, this definition of weight coincides with that given earlier for a standard Weyl tableau.

The constraint in a standard Weyl tableau that each row (column) should comprise a set of nondecreasing (strictly increasing) nonnegative integers is realized in a Gel'fand pattern by the 'geometrical' rule that the integers (m_{ij}) satisfy the betweenness conditions.

III. *Carrier spaces of the representations of the symmetric group.*

Two important pattern results for the symmetric group S_n are:
a) The set of irreps of S_n is in one-to-one correspondence with the set of partitions $((\lambda))$ of n into n parts; (b) the set of basis vectors of a carrier space of irrep (λ) of S_n is in one-to-one correspondence with the set of standard Young tableaux of shape (λ) having weight $(W) = (1,1,\ldots,1)$. (The number of basis vectors (the number of standard patterns) is then the dimension of the irrep.)

This latter result may, of course, also be expressed in terms of Gel'fand patterns. For example, the irreps of S_3 are enumerated by the partitions of 3 into 3 parts (3 0 0), (2 1 0), and (1 1 1). The standard

Young tableaux of weight (1,1,1) having these shapes, respectively, are

$$\boxed{1}\boxed{2}\boxed{3}\;;\quad \begin{array}{|c|c|}\hline 1 & 2 \\ \hline 3 & \\ \cline{1-1}\end{array}\;,\quad \begin{array}{|c|c|}\hline 1 & 3 \\ \hline 2 & \\ \cline{1-1}\end{array}\;;\quad \begin{array}{|c|}\hline 1 \\ \hline 2 \\ \hline 3 \\ \hline\end{array}\quad . \tag{1o}$$

Thus, the irreps (3 0 0), (2 1 0), and (1 1 1) are of dimensions 1, 2 and 1, respectively. These same results are enumerated by the Gel'fand patterns

$$\begin{pmatrix} 3\;0\;0 \\ 2\;0 \\ 1 \end{pmatrix} ; \begin{pmatrix} 2\;1\;0 \\ 2\;0 \\ 1 \end{pmatrix} , \begin{pmatrix} 2\;1\;0 \\ 1\;1 \\ 1 \end{pmatrix} ; \begin{pmatrix} 1\;1\;1 \\ 1\;1 \\ 1 \end{pmatrix} . \tag{11}$$

The standard Young tableaux for S_n are often enumerated by another indexing scheme -- the *Yamanouchi symbol* (5)

$$(y) = (y_1, y_2, \ldots, y_n) \quad . \tag{12}$$

Here y_{n-j+1} is the positive integer equal to the row in which j appears in a given standard Young tableau of shape (λ) and weight (1,1,...,1). y_{n-j+1} is also the position (counting from the left) in which 1 occurs in the set of differences

$$m_{1j}-m_{1j-1}, m_{2j}-m_{2j-1}, \ldots, m_{j-1j}-m_{j-1j-1}, m_{jj} \tag{13}$$

formed from the entries in the corresponding Gel'fand pattern $(m_{1n} \ldots m_{nn}) = (\lambda_1 \ldots \lambda_n)$ having weight (1,1,...,1). For example, the Yanouchi symbols for the Young tableaux (1o) (and Gel'fand patterns (11), are, respectively,

$$(1,1,1); \quad (2,1,1), \quad (1,2,1); \quad (3,2,1) \quad .$$

IV. *Carrier spaces of the representations of the rotation group.*

The important pattern results for the rotation group (SU(2)) are: (a) The set of irreps of the rotation group is in one-to-one correspondence with the set of partitions (2j 0), j = 0,1/2,1,...; (b) the set of basis vectors of the carrier space of irrep (2j 0) is in one-to-one correspondence with the set of Gel'fand patterns having the partition (2j 0):

$$\begin{pmatrix} 2j & & 0 \\ & j+m & \end{pmatrix} \quad , m = -j, -j+1, \ldots, j \quad . \tag{14}$$

(Observe that the betweenness rule embodies in a natural way the fact that the projection quantum number m runs over the values: $m=-j,\ldots,j$.)

The Weyl tableau corresponding to the Gel'fand pattern (14) is the one-rowed pattern

$$\underbrace{\boxed{1}\,\boxed{1}\,\boxed{1}\;\cdots\;\boxed{1}}_{j+m}\,\underbrace{\boxed{2}\,\boxed{2}\;\cdots\;\boxed{2}}_{j-m} \tag{15}$$

The notation above for SU(2) is a special case of U(2) for which we now give an explicit construction of the basis vectors in terms of boson operators.

The standard Weyl tableau of two rows corresponding to the Gel'fand pattern

$$\begin{pmatrix} m_{12} & & m_{22} \\ & m_{11} & \end{pmatrix} \; , \text{ where } m_{12} \geqslant m_{11} \geqslant m_{22} \quad , \tag{16}$$

is

$$\begin{array}{l} \overbrace{\boxed{1}\,\boxed{1}\;\cdots\;\boxed{1}}^{m_{22}}\,\overbrace{\boxed{1}\,\boxed{1}\;\cdots\;\boxed{1}}^{m_{11}-m_{22}}\,\overbrace{\boxed{2}\,\boxed{2}\;\cdots\;\boxed{2}}^{m_{12}-m_{11}} \\ \boxed{2}\,\boxed{2}\;\cdots\;\boxed{2} \end{array} \tag{17}$$

A mapping from Weyl tableaux to bosons is given by

$$\begin{array}{l} 1 \rightarrow \begin{pmatrix} 1 & & 0 \\ & 1 & \end{pmatrix} \rightarrow a_1^1 \quad , \\ 2 \rightarrow \begin{pmatrix} 1 & & 0 \\ & 0 & \end{pmatrix} \rightarrow a_2^1 \quad , \\ \begin{smallmatrix}1\\2\end{smallmatrix} \rightarrow \begin{pmatrix} 1 & & 1 \\ & 1 & \end{pmatrix} \rightarrow \quad \det\begin{pmatrix} a_1^1 & a_1^2 \\ a_2^1 & a_2^2 \end{pmatrix} \equiv a_{12}^{12} \end{array} \tag{18}$$

(The Weyl tableau $\begin{array}{|c|}\hline 1 \\ \hline 2 \\ \hline \end{array}$ corresponds to antisymmetrized bosons made up of two independent bosons a_i^1 and a_i^2 $(i=1,2,)$.)

Using the correspondence (18), we obtain the following boson state vector, corresponding to the Gel'fand pattern (16) and the Weyl tableau (17):

$$\left|\begin{pmatrix} m_{12} & & m_{22} \\ & m_{11} & \end{pmatrix}\right\rangle = M^{-1/2}\cdot(a^{12}_{12})^{m_{22}}(a^{1}_{1})^{m_{11}-m_{22}}(a^{1}_{2})^{m_{12}-m_{11}}|0\rangle \; . \qquad (19)$$

where the normalization factor is given by:

$$M=\frac{(m_{12}+1)!\,(m_{11}-m_{22})!\,(m_{12}-m_{11})!\,(m_{22})!}{(m_{12}-m_{22}+1)!} \; . \qquad (2o)$$

The angular momentum labels for the states (19) are

$$j=\frac{m_{12}-m_{22}}{2}\;, \qquad m=m_{11}-\frac{m_{12}+m_{22}}{2} \; . \qquad (21)$$

(The $2m_{22}$ anti-symmetric (paired) bosons are inert as far as angular momentum is concerned, that is, a^{12}_{12} is invariant under unitary unimodular transformations.)

V. *Double tableaux and the rotation matrices.*

A closer inspection of the basis vectors (19) reveals that the Weyl tableau (17) has been used to assign the *subscripts* to the bosons. One sees, in fact, that the superscript assignment originates from the Weyl tableau

$$\begin{array}{c} \overbrace{\begin{array}{|c|c|c|c|}\hline 1 & 1 & \cdots & 1 \\ \hline 2 & 2 & \cdots & 2 \\ \hline\end{array}}^{m_{22}} \!\!\!\begin{array}{c} \overbrace{\begin{array}{|c|c|c|c|}\hline 1 & 1 & \cdots & 1 \\ \hline\end{array}}^{m_{12}-m_{22}} \\ \end{array}\end{array} \qquad (22)$$

corresponding to the maximal Gel'fand pattern

$$\begin{pmatrix} m_{12} & & m_{22} \\ & m_{12} & \end{pmatrix} \; . \qquad (23)$$

A more descriptive notation for the state vector (19) uses a *double Weyl tableau* or a *double Gel'fand pattern*:

$$\left|\begin{array}{|c|c|c|c|c|c|c|c|c|}\hline 1 & \cdots & 1 & 1 & \cdots & 1 & 2 & \cdots & 2 \\ \hline 2 & \cdots & 2 \\ \cline{1-3}\end{array}\;\middle|\;\begin{array}{|c|c|c|c|c|c|c|c|c|}\hline 1 & \cdots & 1 & 1 & \cdots & 1 & 1 & \cdots & 1 \\ \hline 2 & \cdots & 2 \\ \cline{1-3}\end{array}\right\rangle \qquad (24)$$

$$= \left|\begin{pmatrix} & m_{12} & \\ m_{12} & & m_{22} \\ & m_{11} & \end{pmatrix}\right\rangle = M^{-1/2}(a^{12}_{12})^{m_{22}}(a^1_1)^{m_{11}-m_{22}}(a^1_2)^{m_{12}-m_{11}}|0\rangle$$

where we observe that

(*i*) the Young frames have the *same shape;*

(*ii*) by convention the second Gel'fand pattern (23) is inverted over the first one (16) in order to depict explicitly the shared labels ($m_{12}m_{22}$) giving the common shape of the Young frame;

(*iii*) the mapping from the double Weyl tableau to bosons is obtained by pairing off the columns occurring in the *same positions* in the two Weyl tableaux

$$\left\{\begin{array}{|c|}\hline 1\\\hline 2\\\hline\end{array}\ ,\ \begin{array}{|c|}\hline 1\\\hline 2\\\hline\end{array}\right\} \rightarrow a^{12}_{12} \qquad \left\{\boxed{i}\ ,\ \boxed{j}\right\} \rightarrow a^j_i \quad , \quad i,j = 1,2. \tag{25}$$

(In the patterns in (24) the column pair

$$\left\{\begin{array}{|c|}\hline 1\\\hline 2\\\hline\end{array}\ ,\ \begin{array}{|c|}\hline 1\\\hline 2\\\hline\end{array}\right\}$$

occurs m_{22} times; the column pair $\{\boxed{1}\ ,\ \boxed{1}\}$ occurs $m_{11}-m_{22}$ times, and the column pair $\{\boxed{1}\ ,\ \boxed{2}\}$ occurs $m_{12}-m_{11}$ times.)

The significance of rewriting Eq. (19) in the form of Eq. (24) is that one now recognizes that the latter result generalizes: *The Weyl tableau in the second position* (the upper Gel'fand pattern) *may be taken to be any standard tableau corresponding to the shape* ($m_{12}\ m_{22}$). The mapping (25) then assigns a definite state vector (boson polynomial) to each pair of standard Weyl tableaux of the same shape.

The method outlined above for associating boson polynomials to double standard tableaux is the natural extension of Eq. (24) and is of interest in its own right (cf. Doubilet, Rota and Stein (6)), but it leads to non-orthogonal boson state vectors, except for the special case (24) (cf. Eq. (35) below). We therefore develop an alternative method, used primarily by physicists, which utilizes repeated application of a lowering operator,

$$E^{21} = \sum_{i=1}^{2} a^2_i\ \bar{a}^1_i \quad , \tag{26}$$

to the vector (24), thereby generating orthonormal boson state vectors. These orthonormal vectors may be expressed in an elegant combinatoric form:

$$\left|\begin{pmatrix} & m'_{11} & \\ m_{12} & & m_{22} \\ & m_{11} & \end{pmatrix}\right\rangle = M^{-1/2}\, B\begin{pmatrix} & m'_{11} & \\ m_{12} & & m_{22} \\ & m_{11} & \end{pmatrix}(A)|\,0\,\rangle\ , \quad (27)$$

where

$$B\begin{pmatrix} & m'_{11} & \\ m_{12} & & m_{22} \\ & m_{11} & \end{pmatrix}(A) = [w_1!w_2!w'_1!w'_2!]^{1/2}\,(a^{12}_{12})^{m_{22}}\ \times$$

$$\sum_{\boxed{\alpha}} \frac{(a^1_1)^{\alpha^1_1}\ (a^1_2)^{\alpha^1_2}\ (a^2_1)^{\alpha^2_1}\ (a^2_2)^{\alpha^2_2}}{(\alpha^1_1)!\ (\alpha^1_2)!\ (\alpha^2_1)!\ (\alpha^2_2)!}\ , \quad (28)$$

in which (W) and W') are, respectively, weights of the Gel'fand patterns

$$\begin{pmatrix} m_{12} & & m_{22} \\ & m_{11} & \end{pmatrix} \quad \text{and} \quad \begin{pmatrix} m_{12} & & m_{22} \\ & m'_{11} & \end{pmatrix} \quad (29)$$

and the summation is over all nonnegative integers α^j_i such that the matrix α has the fixed row and column sums given by (W) and (W'), that is,

$$\boxed{\alpha} = \begin{array}{|cc|c} \alpha^1_1 & \alpha^2_1 & w_1 \\ \alpha^1_2 & \alpha^2_2 & w_2 \\ \hline w^1_1 & w^1_2 & \end{array} \quad (30)$$

$(\alpha^1_1 + \alpha^2_1 = w_1,\ \alpha^1_2 + \alpha^2_2 = w_2,\ \alpha^1_1 + \alpha^1_2 = w^1_1,\ \alpha^2_1 + \alpha^2_2 = w^1_2.)$

Observe that while the double Gel'fand patterns in Eq. (27) are in one-to-one correspondence with the double standard Weyl tableaux, we no longer have a simple rule for reading off the general form (28).

We will not give the details here of the derivation of Eqs. (27) and (28), but let us note several important properties of the *double Gel'fand pattern polynomials* (28):

(*i*) The set of double Gel'fand pattern polynomials of weight (W,W') is a (linearly independent) basis of the vector space spanned by all monomials in the bosons (a_i^j) which contain w_i occurences of the subscript i and w'_j occurrences of the superscript j.

(*ii*) The set of double Gel'fand pattern polynomials corresponding to all partitions (m) of the nonnegative integer N is a basis of the vector space of homogeneous polynomials of degree N in the bosons $\{a_i^j\}$.

(*iii*) The matrix $B^{(m)}(A)$ having element in row m_{11} $(m_{11}=m_{12},\cdots,m_{22})$ and column m'_{11} $(m'_{11}=m_{12},\cdots,m_{22})$ given by the boson polynomials (28) is a unitary irreducible representation of the group U(2) when the matrix A is replaced by a unitary 2x2 matrix.

(*iv*) If we replace the bosons a_i^j in Eq. (28) by the elements u_i^j of a 2x2 unitary *unimodular* matrix U, we obtain the (unitary) irreducible representations of SU(2) (rotation matrices):

$$D^j_{m'm}(U) = B\begin{pmatrix} & j+m' & \\ 2j & & o \\ & j+m & \end{pmatrix}(U) \qquad (31)$$

VI. *The general boson polynomials of U(n)*

Let us turn now to the description of the U(n) boson polynomials stating some of their important properties. There is a vast literature on this subject (cf. Ref. 7-25 and references therein). Our presentation is based on results which may be found in Refs. 6, 9, 1o, 11, 14, 17, 18, and 23, to which we refer for further details and proofs. We first sketch the relationship of the U(n) boson polynomials to double standard tableaux. Consider the double standard Weyl tableau of shape $(\lambda) = (\lambda_1\lambda_2\cdots\lambda_n)$:

$$\left(\begin{array}{|c|c|c|c|c|}\hline i_{11} & i_{12} & \cdots & & i_{12_1} \\ \hline i_{21} & i_{22} & \cdots & i_{2\lambda_2} & \\ \cline{1-4} \vdots & & & & \\ \hline i_{n1} & i_{n2} & \cdots & i_{n\lambda_n} & \\ \cline{1-4} \end{array} \;\middle|\; \begin{array}{|c|c|c|c|c|}\hline j_{11} & j_{12} & \cdots & & j_{1\lambda_1} \\ \hline j_{21} & j_{22} & \cdots & j_{2\lambda_2} & \\ \cline{1-4} \vdots & & & & \\ \hline j_{n1} & j_{n2} & \cdots & j_{n\lambda_n} & \\ \cline{1-4} \end{array}\right) \qquad (32)$$

Alternatively, this double standard tableau may be denoted by the double Gel'fand pattern

$$\begin{pmatrix} (m') \\ [m] \\ (m) \end{pmatrix} \qquad (33)$$

where the left and right tableaux in (32) correspond, respectively, to the upper and lower Gel'fand patterns in (33).

With each pair of columns in corresponding positions in the left and right patterns of the double standard Weyl pattern (32), we now associate a determinantal boson by the rule

$$\left\{\begin{array}{cc} i_{1k} & j_{1k} \\ i_{2k} & j_{2k} \\ \vdots & \vdots \\ i_{\lambda'_k k} & j_{\lambda'_k k} \end{array}\right\} \rightarrow a^{j_{1k}\cdots j_{\lambda'_k k}}_{i_{1_k}\cdots i_{\lambda'_k k}} \qquad ,$$

where

$$a^{j_1\cdots j_k}_{i_1\cdots i_k} \equiv \det \begin{pmatrix} a^{j_1}_{i_1} & \cdots a^{j_k}_{i_1} \\ \vdots & \vdots \\ a^{j_1}_{i_k} & \cdots a^{j_k}_{i_k} \end{pmatrix} \qquad (34)$$

A boson polynomial corresponding to a double standard Weyl tableau is defined as the product of the determinantal bosons (34) taken over all columns $1,2,\ldots,\lambda_1$ of the frame. Using the double Gel'fand patterns to denote the polynomials, we have:

$$P\begin{pmatrix}(m')\\ [m]\\ (m)\end{pmatrix}(A) = \prod_{k=1}^{\lambda_1} a_{i_{1k}\ \cdots\ i_{\lambda'_k k}}^{j_{1k}\ \cdots\ j_{\lambda'_k k}} \tag{35}$$

We note two special cases of Eq. (35)

$$P\begin{pmatrix}(m')\\ [m]\\ (max)\end{pmatrix}(A) = \prod_{k=1}^{n}\begin{pmatrix}a_{12\ \ldots k}^{12\ \ldots k}\end{pmatrix}^{m_{kn}-m_{k+1n}} \tag{36}$$

$$P\begin{pmatrix}(max)\\ [m]\\ (semi\text{-}max)\end{pmatrix}(A) = \prod_{k=1}^{n-1}\begin{pmatrix}a_{12\ \ldots k}^{12\ \ldots k}\end{pmatrix}^{m_{kn-1}-m_{k+1n}}$$

$$\times \prod_{k-1}^{n}\begin{pmatrix}a_{12\ \ldots k-1n}^{12\ \ldots k-1k}\end{pmatrix}^{m_{kn}-m_{kn-1}} \tag{37}$$

where $m_{ij} = 0$ for $i > j$, $a_{12\ldots k-1n}^{12\ldots k-1k} = a_n^k$ for $k = 1$, and special pattern notations have been introduced:

$$\begin{pmatrix}[\dot{m}]\\ (max)\end{pmatrix} = \begin{pmatrix} m_{1n} & m_{2n} & \cdots & m_{n-1n} & m_{nn} \\ \ddots & & \ddots & & \\ & m_{1n} & & m_{2n} & \\ & & m_{1n} & & \end{pmatrix}, \tag{38}$$

$$\begin{pmatrix}m\\ semi\text{-}max\\ (max)\end{pmatrix} \equiv \begin{pmatrix} m_{1n} & m_{2n} & \cdots & m_{n-1n} & m_{nn} \\ m_{1n-1} & m_{2n-1} & \cdots & m_{n-1n-1} & \\ & & (max) & & \end{pmatrix}. \tag{39}$$

The weight (W,W') or content of the double standard tableau (32) and of the double Gel'fand pattern (33) is defined to be (cf. Eqs. (1) and (9))

$$(W,W') = (w_1,\dots,w_n,\ w_1',\dots,w_n') \quad , \qquad (4o)$$

where (W) and (W') are, respectively, the weights of the left and right standard tableaux (upper and lower Gel'fand patterns).

As noted earlier the boson state vectors corresponding to the polynomials (35) are not, in general, orthogonal (cf. Eq. (44) - (46) below), and the main emphasis in physics has been on the construction of *orthonormal basis vectors* denoted in the double Gel'fand pattern notation by

$$\left| \begin{pmatrix} (m') \\ [m] \\ (m) \end{pmatrix} > \right\rangle = \left[M([m])\right]^{-1/2} B \begin{pmatrix} (m') \\ m \\ (m) \end{pmatrix} (A) |\ 0 > \qquad (41)$$

where

$$M([m]) = H^{[m]} = \prod_{i=1}^{n} p_{ij}! / \prod_{i<j} (p_{in} - p_{jn}) \qquad (42)$$

in which

$$p_{in} = m_{in} + n-i \ , \quad (p_{in} \text{ is called a "partial hook"}). \qquad (43)$$

The boson polynomials

$$B \begin{pmatrix} (m') \\ [m] \\ (m) \end{pmatrix} (A) \qquad (44)$$

occurring in Eq. (41) and the double tableau polynomials

$$P \begin{pmatrix} (m') \\ [m] \\ (m) \end{pmatrix} (A) \qquad (45)$$

span the same vector spaces. However, only for the patterns

$$\begin{pmatrix} (max) \\ [m] \\ (max) \end{pmatrix} , \quad \begin{pmatrix} (max) \\ [m] \\ (semi\text{-}max) \end{pmatrix} , \quad \begin{pmatrix} (semi\text{-}max) \\ [m] \\ (max) \end{pmatrix} \qquad (46)$$

do the polynomials agree (up to a normalization factor).

We will now state the form of the boson polynomials (44) referring to Refs. 11, 13, 17, and 23 for a discussion of the properties which characterize these orthonormal forms and for the derivations of the results below.

We begin with the statement of the simplest polynomials which are those corresponding to a Young frame having 1 row with p boxes so that $[m] = [p0\ldots0] = [p\dot{0}]$:

$$B\begin{pmatrix}(m')\\ [p\,\dot{0}]\\ (m)\end{pmatrix}(A) = \left[\prod_{i=1}^{n}(w_i)!(w_i')!\right]^{1/2} \times \sum_{\boxed{\alpha}} \prod_{j=1}^{n} (\alpha_i^j)^{\alpha_i^j} / (\alpha_i^j)! \tag{47}$$

where (W) and (W') denote the *weights* of the lower and upper Gel'fand patterns, respectively, and $\boxed{\alpha}$ denotes the following square matrix of nonnegative integers with constraints on the sums of the entries in the rows and columns:

$$\boxed{\alpha} = \begin{array}{|cccc|c}\hline \alpha_1^1 & \alpha_1^2 & \cdots & \alpha_1^n & w_1\\ \alpha_2^1 & \alpha_2^2 & \cdots & \alpha_2^n & w_2\\ & \vdots & & & \\ \alpha_n^1 & \alpha_n^2 & \cdots & \alpha_n^n & w_n\\ \hline w_1' & w_2' & & w_n' & \end{array} \tag{48}$$

The symbols $w_i(w'_j)$ written to the right of row i (below column j) designate that the entries in row i (column j) are constrained to add to $w_i(w_j')$. The sum over $\boxed{\alpha}$ in Eq. (47) is to be taken over all nonnegative integers α_i^j (for i, j = 1,2,...,n) which satisfy these constraints.

The general result has a form similar to Eq. (47):

$$B\begin{pmatrix}(m')\\ [m]\\ (m)\end{pmatrix}(A) = M^{1/2}(\ m\ )\sum_{\alpha} C\begin{pmatrix}(m')\\ [m]\\ (m)\end{pmatrix}(\alpha) \times \prod_{i,j=1}^{n} (\alpha_i^j)^{\alpha_i^j}/[(\alpha_k^j)!]^{1/2}; \quad (49)$$

where the coefficients C in this result are given by

$$C\begin{pmatrix}(m')\\ [m]\\ (m)\end{pmatrix}(\alpha) = \left\langle \begin{pmatrix}[m]\\ (m')\end{pmatrix} \middle| \left\langle [w \begin{matrix}(\Gamma_n)\\ \\ (\alpha_n)\end{matrix} 0]\right\rangle \cdots \left\langle [w \begin{matrix}(\Gamma_2)\\ \\ (\alpha_2)\end{matrix} 0]\right\rangle \left\langle [w \begin{matrix}(\Gamma_1)\\ \\ (\alpha_1)\end{matrix} 0]\right\rangle \middle| 0\right\rangle \quad (5o)$$

in which $\begin{pmatrix}[w_i & \dot{0}]\\ & \alpha_i\end{pmatrix}$ denotes the Gel'fand pattern

$$\begin{pmatrix}[w_i\ \dot{0}]\\ (\alpha_i)\end{pmatrix} = \begin{pmatrix} \alpha_i^1+\alpha_i^2+\ldots+\alpha_i^n & & 0\ldots 0\\ \ddots & & \ddots\\ & \alpha_i^1+\alpha_i^2 & 0\\ & & \alpha_i^1 \end{pmatrix} \quad (51)$$

where (Γ_k) is the operator pattern which is uniquely determined by the Δ pattern

$$[\Delta(\Gamma_k)] = [m_{1k}m_{2k}\cdots m_{kk}\dot{0}] - [m_{1k-1}m_{2k-1}\cdots m_{k-1k-1}\dot{0}] \quad . \quad (52)$$

We can not go into an explanation here of the general structure of the coefficients (5o), but we will explain how they are calculated for the special case of interest for S_n. It is sufficient here to note that the general coefficients (5o) are explicitly known.

We complete this general discussion with several observations on the properties of the boson polynomial (49): *The important properties (i) and (ii) noted earlier (end of Sec. V) apply as stated to the double Gel'fand pattern polynomials*

$$B\begin{pmatrix}(m')\\ [m]\\ (m)\end{pmatrix}(A) \quad .$$

Property (iii) also generalizes to the group U(n), *where the rows and columns of the matrix* $B^{(m)}$ (A) *are now to be enumerated by the* U(n-1) *Gel'fand patterns* ((m),(m')). (Similar statements also apply to the polynomials (35).) *Finally, we have also the transformation property under the combined left and right translations of the boson matrix,*

$$A \to U\,A\,V, \quad U, V \in U(n), \tag{53}$$

given by

$$B\begin{pmatrix}(m')\\ [m]\\ (m)\end{pmatrix}(\tilde{U}AV) = \sum_{(\mu)\,(\mu')} D^{[m]}_{(\mu)\,(m)}(U)\,D^{[m]}_{(\mu')\,(m')}(V)\,B\begin{pmatrix}(\mu')\\ [m]\\ (m)\end{pmatrix}(A), \tag{54}$$

where denotes matrix transposition, and

$$\{D^{m}(U)\,|\,U \in U(n)\} \tag{55}$$

is the (unitary) matrix representation of U(n) *obtained by the identification*

$$D^{[m]}_{(m)\,(m')}(U) = B\begin{pmatrix}(m')\\ [m]\\ (m)\end{pmatrix}(U)\ . \tag{56}$$

VII . *The Young-Yamanouchi real, proper orthogonal irreducible representations of* S_n.

Let us begin by considering the Cayley n x n permutation representation of S_n. For this one lets P denote a pernumation by the rule:

$$P=\begin{pmatrix}1 & 2 & \dots & n\\ j_1 & j_2 & \dots & i_n\end{pmatrix}\ . \tag{57}$$

Then the correspondence

$$P \to [e_{i_1} e_{i_2} \cdots e_{i_n}] \equiv I_P\ , \tag{58}$$

where e_i denotes a unit column vector with 1 in row i and zeroes elsewhere - is a representation of S_n by nxn matrices.

Since the general boson polynomial admits of an interpretation of the argument A by an nxn indeterminate, it is a well-defined operation to replace A by I_P, in Eq. (49). One obtains

$$B \begin{pmatrix} (m') \\ [m] \\ (m) \end{pmatrix} (I_P)$$

$$= [M([m])^{1/2} \delta_{w_1' w_{i_1}} \delta_{w_2' w_{i_2}} \cdots \delta_{w_n' w_{i_n}} \; C \begin{pmatrix} (m') \\ [m] \\ (m) \end{pmatrix} (a_P) \quad , \tag{59}$$

where a_P denotes the nxn numerical array

$$(a_P) = [w_{i_1} e_{i_1}, w_{i_2} e_{i_2}, \ldots, w_{i_n} e_{i_n}] \quad . \tag{6o}$$

Let us next specialize to representations having labels [m] which are partitions of n, and at the same time *restrict the two Gel'fand patterns* (m) *and* (m') *such that the weights* $[W]=[W'] =[\dot{1}]$. It follows at once from Eq. (59) that these special boson polynomials take the form:

$$B \begin{pmatrix} (m') \\ m \\ (m) \end{pmatrix} (I_P) = [M([m])]^{1/2} \; C \begin{pmatrix} (m') \\ [m] \\ (m) \end{pmatrix} (I_P) \quad . \tag{61}$$

It is useful to give a special notation to these objects; let us define

$$D^{[m]}_{(m),(m')}(P) = B \begin{pmatrix} (m') \\ [m] \\ (m) \end{pmatrix} (I_P) \quad . \tag{62}$$

Then

$$\{ D^{[m]}(P) | P \in S_n \}$$

is an irreducible real, orthogonal representation of S_n.

Consider now the specific form taken by the matrix elements of these irreps. From Eq. (5o) we obtain

$$D^{[m]}_{(m),(m')}(P)=\left[n!/\dim[m]\right]^{1/2} \times \left\langle \binom{[m]}{(m)} \middle| \left\langle [1 \overset{\gamma_n}{\underset{i_n}{}} \dot{0}] \right\rangle \cdots \left\langle [1 \overset{\gamma_2}{\underset{i_2}{}} \dot{0}] \right\rangle \left\langle [1 \overset{\gamma_1}{\underset{i_1}{}} 0] \right\rangle \middle| \binom{[0]}{(0)} \right\rangle$$

$$\text{for } P = \begin{pmatrix} 1 & 2 & \dots n \\ i_1 & i_2 & \dots i_n \end{pmatrix} , \tag{64}$$

where dim[m] denotes the dimension of the irreducible representation [m] of S_n. We shall now explain in detail the meaning of the quantities appearing in Eq. (64):

(i) The symbol

$$\left\langle [1 \overset{\gamma}{\underset{i}{}} \dot{0}] \right\rangle , \tag{65}$$

denotes a *fundamental Wigner operator* of U(n) (cf. Refs. 9, 17, 18) in which $(\overset{1\ \dot{0}}{\underset{i}{}})$ is an abbreviated notation for the n-rowed Gel'fand pattern which has weight (0 ...0 1 0...0) with the 1 appearing in position i; similarly, $(_1 \overset{\gamma}{} {}_0)$ denotes the inverted Gel'fand pattern which has weight (0...0 1 0...0) with the 1 appearing in position γ. Thus, we have

$$i,\gamma = 1,2,\dots,n \tag{66}$$

in the symbol (65). For example, for n=3, there are 9 fundamental Wigner operators, a typical example being

$$\left\langle 1 \overset{2}{\underset{1}{}} \dot{0} \right\rangle = \left\langle \begin{matrix} & & 0 & & \\ & 1 & & 0 & \\ 1 & & 0 & & 0 \\ & 1 & & 0 & \\ & & 1 & & \end{matrix} \right\rangle . \tag{67}$$

(We will see below that, while upper and lower patterns in Eq. (65) run over the same numerical patterns, the role of the two patterns in the definition of a fundamental Wigner operator (65) are qualitatively different.)

(ii) The sequence in integers

$$(\gamma_n, \gamma_{n-1}, \ldots, \gamma_1) \tag{68}$$

appearing in the upper patterns in Eq. (64) is the Yamanouchi symbol of the Gel'fand pattern

$$\begin{pmatrix} [m] \\ (m') \end{pmatrix} \quad . \tag{69}$$

(Cf. Eqs. (12) amd (13).)

Our remaining task is to define the concept of a fundamental Wigner operator in U(n) and to show how the coefficients in Eq. (64) are calculated.

Let $H^{[m]}$ denote a carrier space for irreducible representation [m] of U(n). Then an orthonormal basis of the space $H^{[m]}$ is:

$$\left\{ |(m)\rangle \quad \begin{matrix} \text{(m) is a Gel'fand pattern of the} \\ \text{Young frame } Y_{[m]} \end{matrix} \right\} \quad . \tag{7o}$$

The fundamental Wigner operator denoted by

$$\left\langle [1 \overset{\tau}{\underset{i}{}} \dot{0}] \right\rangle \tag{71}$$

is a mapping $H^{[m]} \to H^{[m]+\Delta(\tau)}$, where $\Delta(\tau)$ is the weight of the pattern $\begin{pmatrix} 1 & \dot{0} \\ \tau & \end{pmatrix}$. (If $m_{\tau n}+1 < m_{\tau+1,n}$, then $H^{[1n]+\Delta(\tau)}$ contains only the zero vector. The mapping (71) is now defined explicitly by giving its action on each basis vector (7o) of $H^{[m]}$:

$$\left\langle [1 \overset{\tau}{\underset{i}{}} \dot{0}] \right\rangle \left| \begin{matrix} [m] \\ (m) \end{matrix} \right\rangle = \sum_{(m')} \left\langle \begin{matrix} [m]+\Delta(\tau) \\ (m') \end{matrix} \right| \left\langle [1 \overset{\tau}{\underset{i}{}} \dot{0}] \right\rangle \left| \begin{matrix} [m] \\ (m) \end{matrix} \right\rangle \times \left| \begin{matrix} [m]+\Delta(\tau) \\ (m') \end{matrix} \right\rangle , \tag{72}$$

where

$$\left\langle \begin{matrix} [m]+\Delta(\tau) \\ (m') \end{matrix} \right| \left\langle [1 \overset{\tau}{\underset{i}{}} \dot{0}] \right\rangle \left. \begin{matrix} [m] \\ (m) \end{matrix} \right\rangle \tag{73}$$

denotes a real number (matrix element) which we now describe.

For the description of the numbers (73), we require a detailed notation for the entries in the rows of a Gel'fand pattern. We introduce the notation $[m]_k = [m_{1k} \dots m_{kk}]$ for the entries in row k, the notation $[1\ \dot{0}]_k$ for the row vector [1 0...0] of length k, and $\Delta_k\ (\tau_k)$ for the row vector of length k which has 1 in position τ_k ($1 \leqslant \tau_k \leqslant k$) and zeroes elsewhere. In terms of this notation each matrix element (73) may be described in the following manner: Each matrix element (73) is zero unless the Gel'fand pattern

$$\begin{pmatrix} [m] + \Delta(\tau) \\ (m') \end{pmatrix}$$

has the form

$$\begin{pmatrix} [m]_n & + \Delta_n(\tau_n) \\ [m]_{n-1} & + \Delta_{n-1}(\tau_{n-1}) \\ & \vdots \\ [m]_i & + \Delta_i(\tau_i) \\ [m]_{i-1} & \\ \vdots & \\ [m]_1 & \end{pmatrix} \tag{74}$$

where for each prescribed pair, τ and i ($1 \leqslant \tau \leqslant n$, $1 \leqslant i \leqslant n$), the sequence of integers $\tau_n, \tau_{n-1}, \dots, \tau_i$ satisfies

$$\tau_n = \tau \text{ and } 1 \leqslant \tau_k \leqslant k \text{ for } k = n-1, \dots, i. \tag{75}$$

Denoting the Gel'fand pattern (74) by the notation

$$\begin{pmatrix} [m] \\ (m) \end{pmatrix}_{\tau_n \dots \tau_i} , \tag{76}$$

we have the result that each of the nonzero matrix elements (73) factorizes in the following manner:

$$\left\langle \begin{pmatrix} [m] \\ (m) \end{pmatrix}_{\tau_n \dots \tau_i} \middle| \left\langle [1 \overset{\tau}{\underset{i}{}} \dot{0}] \right\rangle \middle| \begin{pmatrix} [m] \\ (m) \end{pmatrix} \right\rangle \tag{77}$$

$$= \prod_{k=i}^{n} \left\langle \begin{matrix} [m]_k + \Delta_k(\tau_k) \\ [m]_{k-1} + \Delta_{k-1}(\tau_{k-1}) \end{matrix} \middle| \left\langle [1 \overset{\tau_k}{\underset{\tau_{k-1}}{}} \dot{0}]_k \right\rangle \middle| \begin{matrix} [m]_k \\ [m]_{k-1} \end{matrix} \right\rangle ,$$

in which, by convention, $\tau_{i-1} = i$ and $\Delta_{i-1}(i) = [\dot{0}]_{i-1}$.

Each of the real numbers

$$\left\langle \begin{matrix} [m]_k+\Delta_k(\tau_k) \\ \\ [m]_{k-1}+\Delta_{k-1}(\tau_{k-1}) \end{matrix} \right| \left\langle \begin{matrix} & \tau_k & \\ [1 & & \dot{0}]_k \\ & \tau_{k-1} & \end{matrix} \right\rangle \left| \begin{matrix} [m]_k \\ [m]_{k-1} \end{matrix} \right\rangle \qquad (78)$$

in the product (77) is calles a *reduced* U(k): U(k-1) *matrix element* and has a very simple interpretation in terms of the *pattern calculus rules* developed in Ref. 7. We state these rules here for the special case required to evaluate the factor (78):

The pattern calculus rules (cf. Ref. 14).

(i) Write out two rows of dots and assign the numerical entries of $\Delta_k(\tau_k)$ and $\Delta_{k-1}(\tau_{k-1})$, as shown:

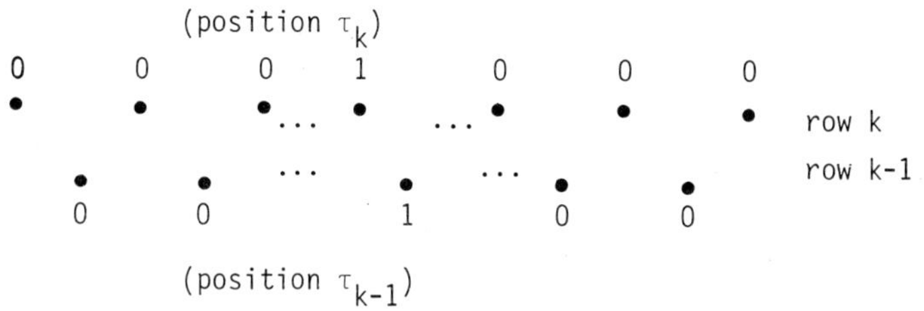

(ii) Draw an arrow between each point labelled by 1 (tail of arrow) to each point labelled by 0 (head of arrow). Once this arrowpattern is drawn, remove the 0's and 1's from the diagram.

(iii) In the arrow-pattern assign the partial hook p_{ik} to point i (i-1,2,...,k from left to right) of row k and the partial hook p_{ik-1} to point i (i=1,2,...,k-1) in row k-1 (p_{ij} m_{ij}+j-i).

(iv) Assign a numerical factor to each arrow in the arrow-pattern using the rule

$$p_{tail} - p_{head} + e_{tail} \quad ,$$

where e_{tail}=1 if the tail of the arrow is on row k-1 and e_{tail}=0 if tail of the arrow is on row k.

(v) Write out the products

N = product of all factors for arrows going *between* rows,

D = product of all factors for arrows going *within* rows.

The reduced U(k):U(k-1) matrix element (78) is then given by

$$S(\tau_{k-1}-\tau_k)\left[\frac{N}{D}\right]^{1/2} \tag{79}$$

where $S(\tau_{k-1}-\tau_k)$ is +1 for $\tau_{k-1} > \tau_k$ and -1 for $\tau_{k-1} < \tau_k$.

Example. for k=3, τ_3=1, τ_2=2 the arrow-pattern is

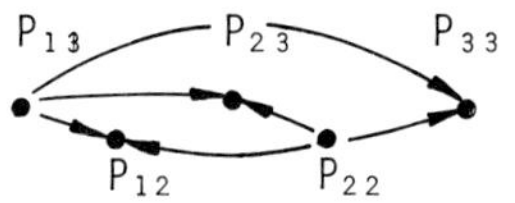

and the reduced matrix element (78) has the value given by

$$\left\langle \begin{matrix} m_{13}+1 & m_{23} & m_{33} \\ m_{12} & m_{22}+1 & \end{matrix} \right| \begin{pmatrix} & & 1 & & \\ & 1 & & 0 & \\ 1 & & 0 & & 0 \\ & 1 & & 0 & \\ & & 0 & & \end{pmatrix} \left| \begin{matrix} m_{13} & m_{23} & m_{33} \\ m_{12} & m_{22} & \end{matrix} \right\rangle$$

$$= - \left[\frac{(p_{13}-p_{12})\ (p_{22}-p_{33}+1)\ (p_{22}-p_{33}+1)}{(p_{13}-p_{23})\ (p_{13}-p_{33})(p_{22}-p_{12}+1)}\right]^{1/2} \tag{80}$$

The result of applying these rules to Eqs. (78) is:

$$S(\tau_{k-1}-\tau_k) \prod_{\substack{s=1 \\ s\neq k}}^{k} \left[\frac{(p_{\tau_{k-1}k-1}-p_{sk}+1)}{(p_{\tau_k k}-p_{sk})} \prod_{\substack{t=1 \\ t\neq\tau_{k-1}}}^{k-1} \frac{(p_{\tau_k k}-p_{tk-1})}{(p_{\tau_{k-1}k-1}-p_{tk-1}+1)}\right]^{1/2} . \tag{81}$$

Remarks. Using the above results from the pattern calculus, Eq. (64) *is a completely explicit general result, giving for each* $P\varepsilon S_n$, *every element of the irreducible matrix representation* $D^{[m]}$ (P).

Thus we have achieved our stated goal of obtaining the real orthogonal S_n irreps in an explicit, non-recursive, way. The techniques we have used, in particular the pattern calculus, seem to be a natural extension of the ideas underlying the concept of a "hook" (due to Nakayama, and to Frame, et al. (26)) as applied in "hook product" of the Hall-Robinson formula (27).

References

1. I.M. Gel'fand and M.L. Tseitlin, Matrix Elements for the unitary groups, Dokl. Akad. Hauk SSSR (195o), 825-828.

2. M. Weyl, Gruppentheorie und Quantenmechanik", Hirzel, Leipzig: 1st ed., 1928; 2nd ed., 1931, translated as "The Theory of Groups and Quantum Mechanics", by M.P. Robertson, Methuen, London, 1931; reissued Dover, New York, 1949.

3. A Young, Quantitative Substitutional Analysis, PLMS (1), I, 33 (19o1), 97-146; II, 34 (19o1), 361-397, PLMS (2) III, 28 (1928), 255-292; IV, 31 (193o), 253-272; V. ibid., 273-288; VI, 34 (1932), 196-23o; VII, 36 (1933), 3o4-368, VIII, 37 (1934), 441-495; IX, 54 (1952), 218-253.

4. H. Weyl, "The Structure and Representations of Continuous Groups", Lectures at the Inst. for Adv. Study, Princeton, N.J., 1934-35 (unpublished). (Notes by R. Brauer.)

5. T. Yamanouchi, On the calculation of atomic energy levels IV, Proc. Phys.-Math. Soc. Japan 18 (1936), 623-64o; On the construction of unitary irreducible representations of the symmetric group, ibid. 19 (1937), 436-45o.

6. P. Doubilet, G.-C. Rota, and J. Stein, On the foundations of combinatorial theory: IX. Combinatorial methods in invariant theory, Studies in Appl. Math., Vol. LIII, No. 3, 1974, 185-216.

7. V. Bargmann and M. Moshinsky, Group theory of harmonic oscillators (I). The collective modes, Nucl. Phys. 18 (196o), 697-712.

8. M. Moshinsky, Bases for the irreducible representations of the unitary groups and some applications, J. Math. Phys. 4 (1963), 1128-1139.

9. G.E. Baird and L.C. Biedenharn, On the representations of semisimple Lie groups. II, J. Math. Phys. 4 (1963), 1449-1466; III, The explicit conjugation operation for SU(n), Ibid, 5 (1964), 1723-173o; IV, A canonical classification for tensor operators in SU_3, Ibid, 5 (1965), 173o-1747.

1o. J.G. Nagel and M. Moshinsky, Operators that raise or lower the irreducible vector spaces of U_{n-1} contained in an irreducible vector space of U_n, J. Math. Phys. 6 (1965), 682-694.

11. J.D. Louck, Group theory of harmonic oscillators in n-dimensional space, J. Math. Phys. 6 (1965), 1786-18o4.

12. M. Moshinsky, Gel'fand states and the irreducible representations of the symmetric group, J. Math. Phys. 7 (1966), 691-698.

13. L.C. Biedenharn, A. Giovannini, and J.D. Louck, Canonical definition of Wigner operators in U_n, J. Math. Phys. 8 (1967), 691-7oo.

14. L.C. Biedenharn and J.D. Louck, A pattern calculus for tensor operators in the unitary groups, Commun. Math. Phys. 8 (1968) 8o-131.

15. P. Kramer and M. Moshinsky, Group theory of harmonic oscillators and nuclear structure, in "Group Theory and its Applications", ed. by E.M. Loebl, Academic Press, New York, 1968, 339-468.

16. M. Ciftan, On the combinatorial structure of state vectors in U(n), II. The generalization of hypergeometric functions on U(n) states, J. Math. Phys. 1o (1969), 1635-1646.

17. J.D. Louck, Recent progress toward a theory of tensor operators in the unitary groups, Am. J. Phys. 38 (197o), 3-42.

18. J.D. Louck and L.C. Biedenharn, Canonical unit adjoint tensor operators in U(n), J. Math. Phys. 11 (197o), 2368-2414.

19. A.C.T. Wu, Structure of the combinatorial generalization of hypergeometric functions for SU(n) states, J. Math. Phys. 12 (1971), 437-44o.

2o. W.J. Holman, On the general boson states of U_n*U_n and Sp_4*Sp_4, Nuovo Cimento 4A (1971), 9o4-931.

21. W.J. Holman and L.C. Biedenharn, "The Representations and Tensor Operators of the Unitary Groups U(n)", in "Group Theory and its Applications", Vol. II, ed. by E.M. Loebl, Academic Press, New York, 1971, 1-73,

22. T.H. Seligman, The Weyl basis of the unitary group U(k), J. Math. Phys. 13 (1972), 876-879.

23. J.D. Louck and L.C. Biedenharn, The structure of the canonical tensor operators in the unitary groups. III. Further developments of the boson polynomials and their implications, J. Math. Phys. 14 (1973), 1336-1357.

24. T.H. Seligman, "Double Coset Decompositions of Finite Groups and the Many-Body Problem", Burg Verlag, A.G., Basel (1975).

25. C.W. Patterson and W.G. Harter, "Canonical symmetrization for the unitary bases. I. Canonical Weyl bases, J. Math. Phys. 17 (1976), 1125-1136; II. Boson and fermion bases, ibid, 17, 1137-1142.

26. J.S. Frame, G. de B. Robinson and R.M. Thrall, "The Hook Graphs of the Symmetric Group", Can. J. Math. 6, 316-324 (1954).

27. G. de B. Robinson, "Representation Theory of the Symmetric Group," University of Toronto Press, 1961.

THE PERMUTATION GROUP AND THE COUPLING OF n SPIN- $\frac{1}{2}$ ANGULAR MOMENTA

J. D. LOUCK†

Group T-7, Theoretical Division
Los Alamos Scientific Laboratory
Los Alamos, New Mexico 87545, USA

L. C. BIEDENHARN*

Physics Department, Duke University
Durham, North Carolina 27706, USA

ABSTRACT

The classic problem of constructing the sharp spin states for n spin- $\frac{1}{2}$ particles by simultaneously classifying the states by their irreducible transformation properties under both SU(2) and S_n is solved explicitly by recognizing that these states are a special case of the boson polynomials of U(n).

†Paper presented by J. D. Louck. Work performed under the auspices of the USERDA
*Sponsored in part by NSF.

I. Introduction

Much of this workshop has dealt with the rôle of the permutation group in elucidating the classical structure of complex molecules. The rôle of the permutation group in the quantum theory of molecules, while certainly not ignored, has perhaps received less emphasis. I wish therefore to discuss the role of the permutation group in a simple but important problem in quantum mechanics.

The problem I have in mind is that of constructing the quantum spin states of a n electron systen such that (i) the states have sharp total spin angular momentum; and (ii) the states transform irreducibly under permutations of the electrons. The importance of this special result in implementing the Pauli principle is well known.

The solution to this problem has, of course, been given by numerous authors, using a variety of techniques (a few of the references are given below, Ref. 1-4). H. Weyl (5) had already given the structure of the general result in 1928, but he did not give the explicit answer. My only excuse in presenting this result anew is to illustrate the novelty and explicitness of results which can be obtained as special cases of the boson polynomials discussed in Prof. Biedenharn's talk. (6). This classic problem is also a nice one for illustrating a cooperative rôle of two group structures: The symmetric group S_n and the quantum mechanical rotation group SU(2).

II. Formulation of the Problem

In order to give a precise statement of the problem, we first summarize the properties of the irreducible representations of SU(2) and S_n.

A. Irreducible representations of the quantum mechanical rotation group.

The Lie algebraic and Hilbert space properties of the total angular momentum $\vec{J} = (J_1, J_2, J_3)$ of a physical system (J_i is the component of $\vec{J}$ relative to an inertial frame of reference) may be summarized as follows:

(1). Each J_i is a linear Hermitian operator on a separable Hilbert space H. The components of $\vec{J}$ satisfy the commutation relations

$$\vec{J} \times \vec{J} = i\,\vec{J}\ . \tag{1}$$

(2). H contains a subspace H_j of dimension $2j+1$, $j \in \{0, 1/2, 1, \ldots\}$, which is invariant and irreducible under the action of J_1, J_2, J_3. The space H_j possesses an orthonormal basis

$$\{|jm\rangle \qquad m = j, j-1, \ldots, -j\} \tag{2}$$

such that the vectors in this basis are simultaneous eigenvectors of the square of total angular momentum, $J^2 = J_1^2 + J_2^2 + J_3^2$, and of J_3, that is,

$$J^2|jm\rangle = j(j+1)\,|jm\rangle\ , \tag{3}$$

$$J^3|jm\rangle = m|jm\rangle\ ; \tag{4}$$

furthermore, the action of J_1 and J_2 on this basis is expressed by

$$(J_1 \pm iJ_2)|jm\rangle = [(j\mp m)(j\pm m+1)]^{1/2}|jm\pm 1\rangle\ . \tag{5}$$

The irreducible representation of SU(2) carried by the space H_j has the following properties.

Let $R(\phi,\hat{n})$ denote a rotation of the Euclidean space R^3 about an axis specified by the unit vector $\hat{n}$ by an angle ϕ. The Hilbert space H_j then undergoes a transformation onto itself which is given up to $\pm$ sign by the unitary operator

$$U = e^{-i\phi\hat{n}\cdot\vec{J}} . \tag{6}$$

In particular, the transformation U of the basis (2) is

$$U|jm\rangle = \sum_{m'} D^j_{m'm}(U)|jm'\rangle , \tag{7}$$

where U denotes the 2x2 unitary matrix

$$U = e^{-i\phi(\hat{n}\cdot\vec{\sigma})/2} = 1\cos\frac{\phi}{2} - i(\hat{n}\cdot\vec{\sigma})\sin\frac{\phi}{2} \tag{8}$$

in which $\vec{\sigma} = (\sigma_1,\sigma_2,\sigma_3)$ denotes the Pauli matrices. $D^j_{m'm}$ denotes a function which maps each $U \in SU(2)$ to the complex numbers and is given explicitly by

$$D^j_{m'm}(u) = [(j+m')!(j-m')!(j+m)!(j-m)!]^{1/2}$$

$$\times \sum_{\boxed{\alpha}} \frac{(u^1_1)^{\alpha^1_1}(u^1_2)^{\alpha^1_2}(u^2_1)^{\alpha^2_1}(u^2_2)^{\alpha^2_2}}{(\alpha^1_1)!\,(\alpha^1_2)!\,(\alpha^2_1)!\,(\alpha^2_2)!} , \tag{9}$$

where

$$U = \begin{pmatrix} u^1_1 & u^2_1 \\ u^1_2 & u^2_2 \end{pmatrix} . \tag{10}$$

The summation is over all nonnegative integers α^j_i which have the fixed row and column sums indicated by the notation

(cf. Ref. 6)

$$\begin{array}{|cc|l} \hline \alpha_1^1 & \alpha_1^2 & j+m' \\ & & \\ \alpha_2^1 & \alpha_2^2 & j-m' \\ \hline \multicolumn{1}{c}{j+m} & \multicolumn{1}{c}{j-m} & \end{array} \quad . \tag{11}$$

Letting $D^j(U)$ denote the $(2j+1)x(2j+1)\cdot$matrix representation of U on the space H, we have: The group of matrices

$$D^j = \{D^j(U) \mid U \varepsilon SU(2)\} \tag{12}$$

is an irreducible unitary representation of SU(2). Furthermore, letting J = 0, 1/2, 1,..., we obtain all the inequivalent irreducible unitary representations of SU(2).

B. Irreducible representations of the symmetric group

A great deal has already been said in this conference about the irreducible representations of the symmetric group, S_n, including an explicit construction of the Yamanouchi real orthogonal representations given in Prof. Biedenharn's talk. We will recall here three basic results:

(i) The irreducible representations of S_n are in one-to-one correspondence with the partitions $[\lambda_1 \lambda_2 \dots \lambda_n]$, $\sum_i \lambda_i = n$, $\lambda_1 \geq \lambda_2 \geq \dots \geq \lambda_n \geq 0$, of n into not more than n nonzero parts. Each such partition $[\lambda]$ also defines a Young frame

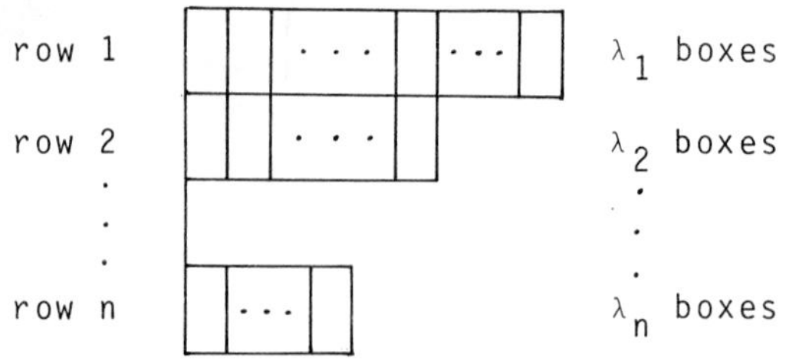

(13)

Thus, an irreducible representation of S_n may be denoted by $\Gamma^{[\lambda]}$ and the set of all (inequivalent) irreducible representations by $\{\Gamma^{[\lambda]} \mid [\lambda]$ is a partition of n }.

(ii) The basis vectors of a linear vector space $V^{[\lambda]}$ which carries an irreducible representation of S_n are in one-to-one correspondence with the set of standard Young tableaux of shape $[\lambda]$. (A standard Young tableaux is a Young frame in which the n boxes have been "filled in" with the integers 1,2,...,n without repitition and such that the sequence of integers appearing in any row or any column is strictly increasing when read from left to right across the row and from top to bottom down the column.)

The Yamanouchi symbol (y) of a standard Young tableau is the sequence of integers $(y) = (y_1, y_2, \ldots, y_n)$ in which y_{n-s+1} is the number of the row in which integer s occurs. Different standard tableaux have different Yamanouchi symbols. Thus, a basis of $V^{[\lambda]}$ may be denoted by

$$\left\{ |[\lambda];(y)\rangle \;\middle|\; \begin{array}{l}(y) \text{ is the Yamanouchi symbol of a standard} \\ \text{Young tableau of shape } [\lambda]\end{array} \right\} \tag{14}$$

(iii) Each permutation $P \varepsilon S_n$ defines a mapping of $V^{[\lambda]}$ onto $V^{[\lambda]}$; hence,

$$P: |[\lambda];(y)\rangle \to \sum_{(y')} \Gamma^{[\lambda]}_{(y')(y)}(P) |[\lambda];(y')\rangle \tag{15}$$

Thus, the irreducible representation of S_n carried by $V^{[\lambda]}$ is

$$\Gamma^{[\lambda]} = \left\{ \Gamma^{[\lambda]}(P) \;\middle|\; P\varepsilon S_n \right\}. \tag{16}$$

C. Coupling of n kinematically independent angular momenta

The Hilbert space for describing the union of two physical systems, when considered as a single physical system, is the tensor product of the Hilbert spaces of the individual systems. The space of interest for the determination of the states of the total angular momentum, $\vec{J} = \sum_{k=1}^{n} \vec{J}(k)$, of n individual physical systems labelled by 1,2,...,n, where system k is in the angular momentum state j_k, is the tensor product space $H_{(j_1 \cdots j_n)}$ defined by

$$H_{(j_1 \cdots j_n)} = H_{j_1} \otimes H_{j_2} \otimes \cdots \otimes H_{j_n} . \tag{17}$$

A basis of this space is

$$\left\{ |j_1 m_1 > \otimes | j_2 m_2 > \otimes \cdots \otimes | j_n m_n > \mid \text{each } m_k = j_k, \ldots -j_k \right\} , \tag{18}$$

and the dimension of the space is

$$\dim H_{(j_1 \cdots j_n)} = \prod_k (2j_k + 1) . \tag{19}$$

Let $i_1, i_2, \ldots, i_n$ denote a rearrangement of the integers $1,2,\ldots,n$. We define the action of the permutation

$$P = \begin{pmatrix} 1 & 2 & \ldots & n \\ i_1 & i_2 & \ldots & i_n \end{pmatrix} \tag{20}$$

on the space $H(j_1 \cdots j_n)$ to be the linear mapping

$$P: \ H(j_1 \cdots j_n) \to H(j_{i_1} \cdots j_{i_n}) \tag{21}$$

defined explicitly by

$$P: \ |j_1 m_1 > \otimes \cdots \otimes |j_n m_n > \to |j_{i_1} m_{i_1} > \otimes \cdots \otimes |j_{i_n} m_{i_n} > . \tag{22}$$

For the problem at hand, we require only the special case $j_1 = \ldots = j_n = 1/2$ and will denote the corresponding tensor product space by H:

$$H = H_{\frac{1}{2}} \otimes \cdots \otimes H_{\frac{1}{2}} \tag{23}$$

with basis

$$\left\{ |\tfrac{1}{2} m_1 > \otimes \cdots \otimes |\tfrac{1}{2} m_n > \mid m_k = \tfrac{1}{2}, -\tfrac{1}{2} \right\} \tag{24}$$

so that

$$\dim \quad H = 2^n \ . \tag{25}$$

In this case, we have

$$P: \ H \to H, \text{ each } P \varepsilon S_n \ . \tag{26}$$

Thus, H is the carrier space of a representation Γ of S_n:

$$\Gamma = \{\Gamma(P) \mid P \varepsilon S_n\} \ , \tag{27}$$

where each $\Gamma(P)$ is a $2^n \times 2^n$ permutation matrix.

The unitary rotation (7) of the single electron states $|\frac{1}{2},\frac{1}{2}>$, $|\frac{1}{2},-\frac{1}{2}>$ takes the form

$$U|\frac{1}{2},\frac{1}{2}> = u^1_1|\frac{1}{2},\frac{1}{2}> + u^1_2|\frac{1}{2},-\frac{1}{2}> \ , \tag{28}$$

$$U|\frac{1}{2},-\frac{1}{2}> = u^2_1|\frac{1}{2},\frac{1}{2}> + u^2_2|\frac{1}{2},-\frac{1}{2}> \ .$$

Under a rotation $R(\phi,\hat{n})$ of the composite n-particle system in R^3, the tensor product space H undergoes the transformation (up to $\pm$ sign) given by

$$U: H \to H \ ,$$

$$U: |\frac{1}{2} m_1 > \otimes \ldots \otimes |\frac{1}{2} m_n > \to U|\frac{1}{2} m_1 > \otimes \ldots \otimes U|\frac{1}{2} m_n > \ . \tag{29}$$

Thus, H is the carrier space of the representation

$$D = \{D(U) = U \otimes \ldots \otimes U \quad U \varepsilon SU(2)\} \tag{30}$$

of SU(2) ($\otimes$ here designates the matrix direct product).

The actions of the operators P and U commute in H and correspondingly the matrix representations given by Eqs. (27) and (30) also commute:

$$\Gamma(P)D(U) = D(U)\Gamma(P), \text{ each } P \varepsilon S_n, \text{ each } U \varepsilon SU(2) \ . \tag{31}$$

Accordingly, H is the carrier space of the representation

$$\{D(U)\Gamma(P)\ |P\varepsilon S_n, U\varepsilon SU(2)\} \tag{32}$$

of the direct product group $SU(2) \times S_n$.

We may now give a precise statement of the problem to be solved: Split the space H into a direct sum of carrier spaces of irreducible representations of $SU(2) \times S_n$ or, equivalently, reduce the representation (32) of $SU(2) \times S_n$ into its irreducible components.

III. Solution to the Problem

Since the announced purpose of this talk was to demonstrate that the complete solution to the problem stated above could be obtained as a special case of the boson polynomials discussed in the previous talk (Ref. 6), we proceed directly to that result.

Consider then the U(n)*U(n) boson state vectors in which the double Gel'fand patterns are specialized in the following manner:

$$\left| \begin{pmatrix} & & (\mu) & & \\ \frac{n}{2}+j & \frac{n}{2}-j & 0 & \cdots & 0 \\ & \frac{n}{2}+j & \frac{n}{2}-j & 0 & \\ & & \frac{n}{2}+j \quad \frac{n}{2}-j & & \\ & & \frac{n}{2}+m & & \end{pmatrix} \right\rangle$$

$$= \sum_{k_1 \cdots k_n}{}_0^1 \left\langle \begin{matrix} 2j & 0 \\ & j+m \end{matrix} \right| \left\langle \begin{matrix} & i_1 & \\ 1 & & 0 \\ & k_n & \end{matrix} \right\rangle \left\langle \begin{matrix} & i_2 & \\ 1 & & 0 \\ & k_{n-1} & \end{matrix} \right\rangle \cdots \left\langle \begin{matrix} & i_n & \\ 1 & & 0 \\ & k_1 & \end{matrix} \right\rangle \left| \begin{matrix} 0 & & 0 \\ & 0 & \end{matrix} \right\rangle$$

$$\cdot\, a^1_{2-k_1}\, a^2_{2-k_2} \cdots a^n_{2-k_n} \left| 0 \right\rangle , \qquad (33)$$

where

(i) (μ) is any lexical Gel'fand pattern with weight $(1,1,\ldots,1)$;

(ii) $(i_1 i_2 \cdots i_n)$ is the sequence of 0's and 1's such that under the identification 0 = 2 the sequence becomes the Yamanouchi symbol of the standard Young tableau of shape $[\frac{n}{2}+j, \frac{n}{2}-j]$ that corresponds to the Gel'fand pattern (μ);

(iii) the symbol $\left| \begin{matrix} 2j & 0 \\ & j+m \end{matrix} \right\rangle$ denotes the basis vector $|jm\rangle$ of standard angular momentum theory;

(iv) the action of each of the four fundamental Wigner operators

$$\left\langle 1 \begin{smallmatrix} i \\ \\ k \end{smallmatrix} 0 \right\rangle \qquad (i,k = 0,1)$$

on an arbitrary basis vector $|jm>$ is obtained from the pattern calculus rules. For convenience, we state these results:

$$\begin{aligned}
\left\langle 1 \begin{smallmatrix} 1 \\ \\ 1 \end{smallmatrix} 0 \right\rangle |jm> &= \left[\frac{j+m+1}{2j+1}\right]^{1/2} \left| j+\frac{1}{2}, m+\frac{1}{2} \right\rangle, \\
\left\langle 1 \begin{smallmatrix} 1 \\ \\ 0 \end{smallmatrix} 0 \right\rangle |jm> &= \left[\frac{j-m+1}{2j+1}\right]^{1/2} \left| j+\frac{1}{2}, m-\frac{1}{2} \right\rangle, \\
\left\langle 1 \begin{smallmatrix} 0 \\ \\ 1 \end{smallmatrix} 1 \right\rangle |jm> &= -\left[\frac{j-m}{2j+1}\right]^{1/2} \left| j-\frac{1}{2}, m+\frac{1}{2} \right\rangle, \\
\left\langle 1 \begin{smallmatrix} 0 \\ \\ 0 \end{smallmatrix} 0 \right\rangle |jm> &= \left[\frac{j+m}{2j+1}\right]^{1/2} \left| j-\frac{1}{2}, m-\frac{1}{2} \right\rangle .
\end{aligned} \qquad (34)$$

Remark. Each i_k in Eq. (33) may assume the values 0 or 1 independently. Quite remarkable, the matrix element of the string of Wigner operators appearing in Eq. (33) is automatically zero unless the sequence $(i_1 i_2 \cdots i_n)$ is the Yamanouchi symbol for (μ) as explained in (ii) above.

Consider next the transformation properties of the boson state vectors (33) (as discussed in Ref. 6).

Under the unitary transformation of the n×n boson matrix $A = (a_i^j)$ given by

$$A \to \tilde{V} A, \; V = \begin{pmatrix} U & 0 \\ 0 & I_{n-2} \end{pmatrix}, \; U \in SU(2) \quad , \qquad (35)$$

we see that the basis vectors (33) undergo the transformation (7).

Consider also the unitary transformation of A given by

$$A \to AI_P \quad , \qquad (36)$$

where I_P is the permutation matrix

$$I_P = [e_{i_1} e_{i_2} \ldots e_{i_n}] \tag{37}$$

in which e_i denotes a unit column vector with 1 in row i and zeroes elsewhere, and P denotes the permutation

$$P = \begin{pmatrix} 1 & 2 \ldots & n \\ i_1 & i_2 \ldots & i_n \end{pmatrix} . \tag{38}$$

Then, from the results of Ref. 6, we find that under the transformation (36) of A, the basis vectors (33) undergo the transformation (15) in which

$$[\lambda] = [\tfrac{n}{2} + j \ \tfrac{n}{2} - j \ 0 \ldots 0] , \tag{39}$$

and (y) is the Yamanouchi symbol corresponding to $(i_1 i_2 \ldots i_n)$ as described under (ii) above. (The irreducible representation $\Gamma^{[\lambda]}$ is then the Yamanouchi real orthogonal representation.)

Thus, the orthonormal basis vectors (33) corresponding to fixed n and j ($0 < j < n/2$), but with $m = j, \ldots, -j$ and (μ) running over all lexical patterns of weight $(1,1,\ldots,1)$, are the basis vectors of a vector space which carries the irreducible representation

$$D^j \otimes \Gamma^{[\frac{n}{2}+j \frac{n}{2}-j]} \tag{40}$$

of $SU(2) \times S_n$.

Observe next that since the boson operators a_i^j, $j = 1,2,\ldots,n$ and $i = 3,4,\ldots,n$ do not occur in the basis vectors (33), we can put $a_i^j = 0$ for $j = 1,\ldots,n$ and $i = 3,\ldots,n$. [The general boson polynomial is then zero unless the lower pattern is of the form occurring in the left-hand side of Eq. (33).] If we now let A denote the $2\times n$ matrix

$$A = \begin{pmatrix} a_1^1 & a_1^2 \ldots a_1^n \\ a_2^1 & a_2^2 \ldots a_2^n \end{pmatrix} , \tag{41}$$

then the transformation of A corresponding to Eqs. (35) and (36) is

$$A \rightarrow \tilde{U}AI_P, \; U\varepsilon SU(2), \; P\varepsilon S_n \; . \tag{42}$$

Comparing this transformation of the boson matrix A with the transformations (26) and (29) of H, we see that the appropriate way to make the transcription from the boson basis vectors (33) to basis vectors of H is to make the replacement

$$a^1_{2-k_1} a^2_{2-k_2} \cdots a^n_{2-k_n} \left| 0 \right\rangle$$

$$\rightarrow \left| \begin{matrix} 1 & 0 \\ & k_1 \end{matrix} \right\rangle \otimes \left| \begin{matrix} 1 & 0 \\ & k_2 \end{matrix} \right\rangle \cdots \otimes \left| \begin{matrix} 1 & 0 \\ & k_n \end{matrix} \right\rangle \tag{43}$$

where

$$\left| \begin{matrix} 1 & & 0 \\ & 1 & \end{matrix} \right\rangle = \left| \tfrac{1}{2}, \tfrac{1}{2} \right\rangle , \quad \left| \begin{matrix} 1 & & 0 \\ & 0 & \end{matrix} \right\rangle = \left| \tfrac{1}{2}, \tfrac{-1}{2} \right\rangle \; . \tag{44}$$

Let us summarize the results obtained above. We introduce the simpler notation $|(i_1 i_2 \cdots i_n) \; ; \; jm >$ for the basis vectors obtained from Eq. (33) by the replacement (43):

$$|(i_1 i_2 \cdots i_n); \, jm \rangle = \sum_{k_1 \ldots k_n}{}_{0}^{1} \left\langle \begin{matrix} 2j & 0 \\ & j{+}m \end{matrix} \right| \left\langle \begin{matrix} & i_1 & \\ 1 & & 0 \\ & k_n & \end{matrix} \right\rangle \cdots \left\langle \begin{matrix} & i_n & \\ 1 & & 0 \\ & k_1 & \end{matrix} \right\rangle \left| \begin{matrix} 0 & & 0 \\ & 0 & \end{matrix} \right\rangle$$

$$\left| \begin{matrix} 1 & 0 \\ & k_1 \end{matrix} \right\rangle \otimes \cdots \otimes \left| \begin{matrix} 1 & 0 \\ & k_n \end{matrix} \right\rangle \; . \tag{45}$$

Then the set of vectors

$$\left\{ \left| (i_1 i_2 \cdots i_n); \, jm \right\rangle \; \middle| \; \begin{matrix} (i_1 i_2 \cdots i_n) \text{ is a Yamanouchi symbol} \\ (0{=}2) \text{of a standard Young tableau of} \\ \text{shape } [\tfrac{n}{2} + j \;\; \tfrac{n}{2} - j \,] ; \; m = j, j-1, \ldots, -j \end{matrix} \right\} \tag{46}$$

is an orthonormal basis of a subspace $H(\frac{n}{2} + j, \frac{n}{2} - j)$ of H which is a carrier space of the irreducible representation

$$D^j \otimes \Gamma^{[\frac{n}{2} + j\ \frac{n}{2} - j]} \tag{47}$$

of $SU(2) \times S_n$. The explicit transformation properties are :

$$U \mid (i_1 \ldots i_n);\ jm > = \sum_{m'} D^j_{m'm}(U) \quad (i_1 \ldots i_n);\ jm'> \quad , \tag{48}$$

$$P{:}\mid (i_1 \ldots i_n);\ jm> \rightarrow \sum_{(i_1' \ldots i_n')} \Gamma^{[\frac{n}{2} + j\ \frac{n}{2} - j]}_{(i_1' \ldots i_n'),(i_1 \ldots i_n)}(P) \mid (i_1' \ldots i_n');jm >. \tag{49}$$

The solution to the problem posed in Section II is now completed with the result

$$H = \sum_j \oplus H\left(\frac{n}{2} + j,\ \frac{n}{2} - j\right) , \tag{50}$$

where the sum is over

$$j = \frac{n}{2}, \frac{n}{2} - 1, \ldots , \frac{1}{2} \text{ or } 0 . \tag{51}$$

Equation (50) may also be proved by appealing to the known properties of the boson polynomials: The set of boson polynomials

$$B \begin{pmatrix} & & (m') & & \\ m_{1n} & m_{2n} & 0 & \cdots & 0 \\ m_{1n} & m_{2n} & & 0 & \\ & m_{1n} & m_{2n} & & \\ & & m_{11} & & \end{pmatrix} (A) \tag{52}$$

corresponding to all allowed patterns (m') and m_{11} and to all partitions $[m_{1n} m_{2n}]$ of N spans the space of all polynomials in the 2n bosons a^j_1, $a^j_2 (j = 1,2,\ldots n)$ which are homogeneous of degree N. For the case at hand, we have $N = n = m_{1n} + m_{2n}$ and $2j = m_{1n} = m_{1n} - m_{2n}$ so that $m_{1n} = \frac{n}{2} + j$ and $m_{2n} = \frac{n}{2} - j$, where all partitions $[m_{1n} m_{2n}]$ of n are obtained by letting j run over $\frac{n}{2}, \frac{n}{2} - 1, \ldots, \frac{1}{2}$ or 0. [Observe that this result is applicable to the space H since the mapping (43) can be reversed.]

One could, of course, also prove Eq. (50) by a dimensionality check, it being necessary then to verify that

$$\sum_j \dim D^j \cdot \dim \Gamma^{[\frac{n}{2}+j,\frac{n}{2}-j]} = 2^n \quad , \tag{53}$$

where

$$\dim D^j = (2j+1) \quad , \tag{54}$$

$$\dim \Gamma^{[\frac{n}{2}+j,\frac{n}{2}-j]} = \frac{2(2j+1)}{2j+2+n} \binom{n}{\frac{n}{2}-j} . \tag{55}$$

Remark. Once one knows the result in the form of Eq. (45), it is certainly not necessary to use anything as powerful as the general $U_n * U_n$ boson polynomials to obtain it, since one may prove the validity of Eq. (45) directly. The boson polynomials, however, yield many useful results when specialized in various ways, and our purpose here was to illustrate one such case of physical interest. We hope that the simplicity of the final result justifies our presenting it here.

In concluding, I would like to give one more notation for the basis vectors (45) which illustrates most vividly the interrelationship between SU(2) and S_n. A similar notation has been employed by Rota and collaborators (7) in their studies in combinatorics. This notation empolys two Young frames of the same shape $[\frac{n}{2}+j\ \frac{n}{2}-j]$:

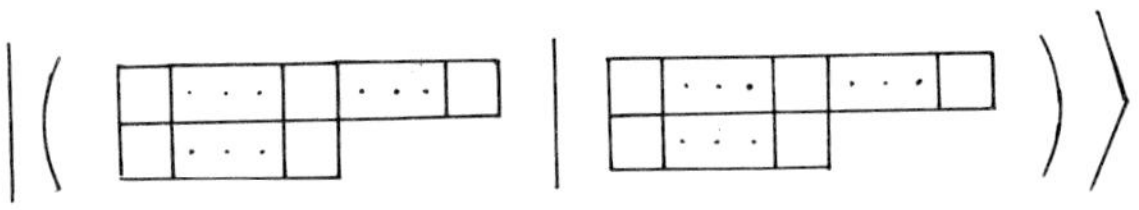

(56)

The Young frame on the right is then filled in with 1,2,...,n to obtain a standard Young tableau; the Young frame on the left is filled in with 1's and 2's to obtain a standard Weyl tableau for SU(2).

REFERENCES

1. P. O. Löwdin, Quantum theory of many-particle systems. III. Extension of the Hartree-Fock scheme to include degenerate systems and correlation effects, Phys. Rev. 91 (1955), 1509-1520.

2. A. Rotenberg, Calculation of exact eigenfunctions of spin and orbital angular momentum using the projection operator method, J. Chem. Phys. 39 (1963), 512-517.

3. R. Paunz, "Alternant Molecular Orbital Method," W. B. Saunders Co., Philadelphia, 1967.

4. J. S. Murty and C. R. Sarma, A method for the construction of orthogonal spin eigenfunctions, Intl. J. Quant. Chem. 9 (1975), 1097-1107.

5. H. Weyl, "Gruppentheorie und Quantenmechanik, Hirzel, Leipzig, 1st ed., 1928; 2nd ed., 1931; translated as "The Theory of Groups and Quantum Mechanics," 1931; reissued, Dover, New York, 1949.

6. L. C. Biedenharn and J. D. Louck, Representations of the symmetric group as special cases of the boson polynomials in U(n) (see previous paper).

7. P. Doubilet, G. C. Rota, J. Stein, On the foundations of combinatorial theory: IX. Combinatorial methods in invariant theory, Studies in Appl. Math., Vol. LIII, No. 3 (1974), 185-216.

THE PERMUTATION GROUP IN ATOMIC STRUCTURE

B. R. Judd
Physics Department, The Johns Hopkins University
Baltimore, Maryland 21218

Symposium on The Permutation Group in Physics and Chemistry,
Centre for Interdisciplinary Research,
University of Bielefeld,
July 1978

The underlying antisymmetry of the fermion states that comprise an atomic shell makes itself felt in its most obvious way in the Slater determinantal product states. The use of these basis states is compared to the methods of Racah, to an approach based on quasiparticles, and to the unitary calculus of Harter. The calculation of the reduced matrix element of a quadrupolar tensor for the ^4P term of g^3 is taken as a running example.

1. INTRODUCTION

In the theory of atomic structure, the permutation group enters at a fundamental level. Electrons are fermions and all electronic states must be totally antisymmetric with respect to the interchange of any two electrons. If, for simplicity, we confine our attention to a single atomic shell, our electrons are labelled by a principal quantum number n, an orbital angular-momentum quantum number ℓ, and two so-called magnetic quantum numbers m_ℓ and m_s. Since the pair $(m_s\ m_\ell)$ can take $2(2\ell + 1)$ possible values, the electronic configuration $(n\ \ell)^N$ comprises $^{4\ell+2}C_N$ states in all. This can easily be quite large: for example, there are 3003 states in $4f^6$. Mixed configurations of the type $(n\ell + n'\ell')^N$ are much more complex. The central problem in the calculation of atomic properties is to organize and arrange the basis functions in the most efficient way. A discussion of the choices open to us forms the subject of this paper.

2. SLATER DETERMINANTS

The most direct form of the basis functions is provided by the Slater determinants. The N electrons are assigned wavefunctions described by the N sets of labels ($n\,\ell\, m_s\; m_\ell$), and then the product is antisymmetrized. The result can be written as a determinant, though for descriptive purposes it is only necessary to list the labels. For example, $\{4^+3^+1^+\}$ is a concise description of a (normalized) determinant of g^3 for which $(m_s\; m_\ell) \equiv (\frac{1}{2}\,4), (\frac{1}{2}\,3), (\frac{1}{2}\;1)$.

The principal defect of the determinantal basis is that, apart from a few special cases, the states are not diagonal with respect to the Coulomb interaction between the electrons. The good quantum numbers for this perturbation are the total spin and total orbital angular momentum, S and L. So, if we want to study a particular SL term, the appropriate linear combination of determinants must be found. The easiest way to do this is to pick M_L (that is, the sum of the individual m_ℓ values) to be L and M_S to be S. Then both operators L_+ ($\equiv L_x + iL_y$) and S_+, when acting on the state $|S\,L\,M_S\,M_L\rangle$, must give a null result. Equations of the type

$$\ell_+|\ell\, m_\ell\rangle = [\ell(\ell+1) - m_\ell(m_\ell+1)]^{\frac{1}{2}}\,|\ell\, m_\ell + 1\rangle$$

serve to determine the ratios of the coefficients of the determinants. For example, we find

$$|g^3\,{}^4P, 3/2, 1\rangle = (4/33)^{\frac{1}{2}}\{4\;1\;{-4}\} - (32/165)^{\frac{1}{2}}\{4\;0\;{-3}\} + (7/33)^{\frac{1}{2}}\{4\;{-1}\;{-2}\} - (3/11)^{\frac{1}{2}}\{3\;2\;{-4}\} + (4/33)^{\frac{1}{2}}\{3\;1\;{-3}\} - (7/330)^{\frac{1}{2}}\{3\;0\;{-2}\} + (1/33)^{\frac{1}{2}}\{2\;1\;{-2}\} + (3/110)^{\frac{1}{2}}\{2\;0\;{-1}\} \quad (1)$$

to within an arbitrary phase factor. (All m_s values are $+\frac{1}{2}$ and have been suppressed.) If, now, we want to find the reduced matrix element of the second-rank unit tensor $\underset{\sim}{U}^{(2)}$ (as we would if the atom were subjected to a quadrupolar electric field), we use the single-electron matrix elements

$$(g\ m_\ell|u_0^{(2)}|g\ m_\ell) = (-1)^{m_\ell}\begin{pmatrix} 4 & 2 & 4 \\ -m_\ell & 0 & m_\ell \end{pmatrix}$$

$$= (13140)^{-\frac{1}{2}}(28,\ 7,\ -8,\ -17,\ -20)$$

for $m_\ell = \pm 4, \pm 3, \pm 2, \pm 1, 0$, and then write (2)

$$({}^4P\|\,U^{(2)}\,\|\,{}^4P) = (30)^{\frac{1}{2}}({}^4P,\ 1\ |U_0^{(2)}|\ {}^4P,\ 1)$$

$$= (30)^{\frac{1}{2}}(13140)^{-1}\ [(4/33)(28 - 17 + 28)$$

$$+ (32/165)(28 - 20 + 7) + \ldots]$$

$$= (189/550)^{\frac{1}{2}}.$$

The factor $(30)^{\frac{1}{2}}$ is the inverse of a 3-j symbol that arises when the Wigner-Eckart theorem is applied in the space of the total orbital angular momentum L.

3. RACAH ALGEBRA

The approach just described becomes extraordinarily tedious if the determinantal expansion is long. Matters are made worse if cross-terms between different determinants have to be considered (as they do for two-body operators such as e^2/r_{ij}). It was Racah (1) who achieved enormous simplifications by introducing recoupling techniques and the coefficients of fractional parentage (cfp). By so doing, the z direction (with its attendant quantum numbers M_S, M_L, m_s, and m_ℓ) is eliminated from all calculations for atoms in which no physically preferred direction exists. The price to be paid is the evaluation of 6-j and 9-j symbols; but, as many of these are now tabulated (or readily available from computer programs), this is no longer a problem. By defining states in terms of their parents, Racah also solved the problem of multiply occurring terms with the same S and L. For these, the techniques used to derive Eq.(1) necessarily fail, since additional quantum numbers are required.

Thus, to obtain the result of Eq.(2), we note that the parents of ^4P of g^3 are ^3F and ^3H of g^2, for which the cfp are (2) $(14/27)^{\frac{1}{2}}$ and $(13/27)^{\frac{1}{2}}$ respectively. The reduced matrix element of $\underset{\sim}{U}^{(2)}$ must be three times that of $\underset{\sim}{u}_3^{(2)}$, where the subscript 3 picks out the third electron of g^3. So

$$|g^3\ {}^4P\rangle = (14/27)^{\frac{1}{2}}|g^2\ {}^3F,\ g_3,\ {}^4P\rangle + (13/27)^{\frac{1}{2}}|g^2\ {}^3H,\ g_3,\ {}^4P\rangle,$$

the commas in the kets indicating a coupling of the third g electron to its parent to produce ^4P. For the matrix element of a tensor acting only on the second part of a coupled system, we have

$$\begin{aligned}({}^4P\,\|\,U^{(2)}\|\,{}^4P) &= 3({}^4P\,\|\,u_3^{(2)}\|\,{}^4P) \\ &= 3(14/27)\cdot 3\begin{Bmatrix}1 & 2 & 1\\ 4 & 3 & 4\end{Bmatrix} + 3(13/27)\cdot 3\begin{Bmatrix}1 & 2 & 1\\ 4 & 5 & 4\end{Bmatrix} \\ &= (189/550)^{\frac{1}{2}},\end{aligned}$$

as before. The 6-j symbols in the above equation can be immediately written down with the aid of the tables of Rotenberg et al.(3)

It is probably fair to say that the techniques of Racah now constitute the standard approach for all atomic configurations of moderate complexity. All cfp are known for $\ell = 1$, 2, and 3, as well as all reduced matrix elements of $U^{(k)}$, thanks to the work of Nielson and Koster.(4) The objections to the method are mainly aesthetic. In spite of the use of the Lie groups, the separation of duplicated terms in f^N is somewhat arbitrary, and little guidance is provided for the terms of the configurations ℓ^N with $\ell > 3$, for which the problem becomes acute. Even in g^3, there are two ^4F terms. But perhaps the most compelling reason for searching for alternative methods lies in the fact that Racah's methods often lead to results of a greater simplicity than one would apparently have any reason to expect. Matrix elements sometimes vanish for no obvious reason, and the eigenvalues of certain operators are occasionally the same for states that seem to have little in common.(2) Most of these phenomena can be understood in terms of quasiparticles, and it is to these that we now turn.

4. QUASIPARTICLES

The possibility of regarding the states for which all $m_s = +\frac{1}{2}$ and those for which all $m_s = -\frac{1}{2}$ as two separate entities has been known for many years. As long ago as 1937, Shudeman (5) coupled the total orbital angular momentum L_A of the first group to that of the second group (L_B) in order to find the SL terms of such complex configurations as g^N, h^N, and i^N. More recently, it was recognized that these two spaces (A and B) can themselves be factored into two parts.(6) To illustrate this, we take a special case. The terms of maximum S (more precisely, those for which all $m_s = +\frac{1}{2}$) of the g shell are as follows:

g^0 and g^9:	S
g^1 and g^8:	G
g^2 and g^7:	P F H K
g^3 and g^6:	P F^2 G H I K M
g^4 and g^5:	S D^2 F G^2 H I^2 K L N

The total list of all L values, for either N even or N odd, runs

$$S^2\ P^2\ D^2\ F^4\ G^4\ H^3\ I^3\ K^3\ L\ M\ N. \qquad (3)$$

This list can be obtained by taking two angular momenta of 2 and 5 units and coupling this pair to an equivalent pair:

$$\begin{aligned} D \times D &= S\ P\ D\ F\ G \\ D \times H &= F\ G\ H\ I\ K \\ H \times D &= F\ G\ H\ I\ K \\ H \times H &= S\ P\ D\ F\ G\ H\ I\ K\ L\ M\ N. \end{aligned} \qquad (4)$$

The sum of the resulting angular momenta coincides with (3).

The origin of the four-fold splitting of the $n\ell$ shell lies in the fact that the four linear quasiparticle combinations

$$\lambda_q^\dagger = (2)^{-\frac{1}{2}} [a_{\frac{1}{2}q}^\dagger + (-1)^{\ell-q} a_{\frac{1}{2}-q}]$$

$$\mu_q^\dagger = (2)^{-\frac{1}{2}} [a_{\frac{1}{2}q}^\dagger - (-1)^{\ell-q} a_{\frac{1}{2}-q}]$$

$$\nu_q^\dagger = (2)^{-\frac{1}{2}} [a_{-\frac{1}{2}q}^\dagger + (-1)^{\ell-q} a_{-\frac{1}{2}-q}] \qquad (5)$$

$$\xi_q^\dagger = (2)^{-\frac{1}{2}} [a_{-\frac{1}{2}q}^\dagger - (-1)^{\ell-q} a_{-\frac{1}{2}-q}]$$

of creation ($a_{m_s m_\ell}^\dagger$) and annihilation ($a_{m_s m_\ell}$) operators anticommute with one another, and hence permit four separate spaces to be constructed. The occurrence of the angular momenta of 2 and 5 for the g shell has a group-theoretical explanation. The vector representation (1000) of R_9 (the rotation group in the nine-dimensional space spanned by the states $|g\, m_\ell>$) yields the single representation G of R_3; but the spinor representation $(\frac{1}{2}\frac{1}{2}\frac{1}{2}\frac{1}{2})$ yields D + H. It is the Kronecker square $(\frac{1}{2}\frac{1}{2}\frac{1}{2}\frac{1}{2})^2$ that provides the representations (0000), (1000), (1100), (1110) and (1111) to which the states of g^0, g^8, g^2, g^6, and g^4 (or, alternatively, g^9, g, g^7, g^3, and g^5) belong.

To illustrate how the quasiparticle method works, we recalculate the reduced matrix element $(g^3\ {}^4P \| U^{(2)} \| g^3\ {}^4P)$. The first step is to express $\underset{\sim}{U}^{(2)}$ in terms of the operators (5). By comparing reduced matrix elements, we find

$$\underset{\sim}{U}^{(2)} = -(2/5)^{\frac{1}{2}} (\underset{\sim}{a}^\dagger \underset{\sim}{a})^{(02)}.$$

With the aid of the inverses of Eqs.(5), we obtain, for an operator acting solely in the spin-up space (space A),

$$\underset{\sim}{U}^{(2)} = -(20)^{-\frac{1}{2}} [(\underset{\sim}{\lambda}^\dagger \underset{\sim}{\mu})^{(2)} + (\mu^\dagger \lambda)^{(2)}]. \qquad (6)$$

(The combinations $(\underset{\sim}{\lambda}^\dagger \underset{\sim}{\lambda})^{(K)}$ and $(\underset{\sim}{\mu}^\dagger \underset{\sim}{\mu})^{(K)}$ vanish identically for K = 2, 4, 6, 8 in virtue of the basic anticommutation relations for fermions.)

As can be seen from the decomposition (4), a P state can be formed from either D × D or H × H. To find the appropriate linear combination, we diagonalize the number operator:

$$N = \sum_{\eta} a_{\eta}^{\dagger} a_{\eta} = 9/2 - (\underset{\sim}{\lambda}^{\dagger} \cdot \underset{\sim}{\mu}^{\dagger}).$$

Standard angular-momentum techniques can be used to handle the last term. The reduced matrix elements for both $\underset{\sim}{\lambda}^{\dagger}$ and $\underset{\sim}{\mu}^{\dagger}$ (denoted generically by $\theta^{\dagger}$) are given for odd N by (6)

$$(D\|\ \theta^{\dagger}\|\ H) = -(55/4)^{\frac{1}{2}},$$

$$(H\|\ \theta^{\dagger}\|\ H) = (143/4)^{\frac{1}{2}}.$$

The matrix for odd N turns out to be

$$\begin{pmatrix} 10/3 & (11/3)^{\frac{1}{2}} \\ (11/3)^{\frac{1}{2}} & 20/3 \end{pmatrix},$$

which has the eigenvalues 7 and 3, as we expect. The eigenfunction for N = 3 allows us to write

$$|g^3\ {}^4P,\ M_S = 3/2> = (11/12)^{\frac{1}{2}}|(D\ D)P> - (1/12)^{\frac{1}{2}}|(H\ H)P>. \qquad (7)$$

The last step involves putting the operator on the right-hand side of Eq.(6) between the state (7) and its adjoint. We need the equations

$$\begin{Bmatrix} \ell_{\lambda} & \ell_{\lambda} & 4 \\ \ell_{\mu} & \ell_{\mu} & 4 \\ 1 & 1 & 2 \end{Bmatrix} = (22/42525)^{\frac{1}{2}},\ (14/18376875)^{\frac{1}{2}},\ (14/323433)^{\frac{1}{2}}$$

for $(\ell_{\lambda}\ \ell_{\mu}) \equiv (2\ 2),(2\ 5),(5\ 5)$ respectively. The result is

$$(g^3\ {}^4P\ \|\ U^{(2)}\ \|\ g^2\ {}^4P) = (1/12)(14/33)^{\frac{1}{2}}\ [121 - 22/5 + 13]$$

$$= (189/550)^{\frac{1}{2}},$$

as before.

Since the above calculation has required a knowledge of 9-j symbols, it may appear that it is not as direct as the previous methods. However, the quasiparticle approach can be used not only for the g shell, but also for the h, i, k, and ℓ shells. Multiplicity problems arise for the first time in the m shell because, for R_{19}, the spinor representation $(\frac{1}{2}\,\frac{1}{2}\,\frac{1}{2}\,\frac{1}{2}\,\frac{1}{2}\,\frac{1}{2}\,\frac{1}{2}\,\frac{1}{2}\,\frac{1}{2})$ contains the angular-momentum states 15/2, 17/2, 21/2, and 27/2 each twice.

5. HARTER'S METHOD

The most recent development in the range of methods open to us is due to Harter.(7) Although it employs the rather recondite language of group theory, it is basically identical to the method of Slater determinants – with, however, one important difference: all the basic states possess S as a good quantum number. The states are defined by Gelfand patterns or their equivalent Young tableaux. For example, the nine m_ℓ states of a g electron are labelled 1, 2,..., 9 for m_ℓ = 4, 3,..., -4. The Slater determinants $\{4^+1^+\text{-}4^+\}$, $\{4^+0^+\text{-}3^+\}$,...,$\{2^+0^+\text{-}1^+\}$ of Eq. (1) become

$$\begin{matrix} 1 & 1 & & 3 \\ 4 & 5 & & 5 \\ 9, & 8, & \ldots, & 6, \end{matrix}$$

the three sets of numbers being arranged as the entries to the tableau

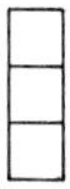

corresponding to a totally antisymmetric orbital state. Since all the Slater determinants of Eq(1) possess M_S = 3/2, for which the only possible value of S is also 3/2, there is no change in the procedure that led to Eq.(2).

The situation changes, however, if S is less than its maximum value. Consider d^3 (a configuration selected for study by Harter and Patterson).(8) The states for which $S = \frac{1}{2}$ are assigned Gelfand states that correspond to entries in the shape

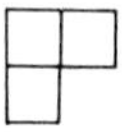

Thus, for M_L = 5 (the maximum), there is one pattern, namely $\begin{smallmatrix}1 & 1\\ 2 & \end{smallmatrix}$, corresponding to the Slater determinant $\{2^+1^+2^-\}$. This can only belong to ^2H. For M_L = 4, there are the two patterns $\begin{smallmatrix}1 & 2\\ 2 & \end{smallmatrix}$ and $\begin{smallmatrix}1 & 1\\ 3 & \end{smallmatrix}$, corresponding to the two Slater determinants $\{2^+1^+1^-\}$ and $\{2^+0^+2^-\}$. One linear combination of these belongs to ^2H, while its orthogonal companion belongs to ^2G. So far, S is a good quantum number for the determinants as well as for the tableaux. However, both ^4F and ^2F exist in d^3, with the result that for M_L = 3, there are four determinantal states, namely

$$\{\,2^+ {-1}^+ 2^-\},\ \{\,2^+1^+0^-\},\ \{\,1^+0^+2^-\},\ \{\,0^+2^+1^-\}.$$

Only the first is an eigenstate of S^2. The linear combination

$$\{2^+1^+0^-\} + \{1^+0^+2^-\} + \{0^+2^+1^-\}$$

corresponds to $S = 3/2$; the remaining two orthogonal combinations belong to $S = \frac{1}{2}$. They can be constructed in an infinity of ways. The two patterns $\begin{smallmatrix}1&3\\2&\end{smallmatrix}$ and $\begin{smallmatrix}1&2\\3&\end{smallmatrix}$ used by Harter correspond to the linear combinations

$$2\{\,2^+1^+0^-\} \ - \ \{\,1^+0^+2^-\} \ - \ \{\,0^+2^+1^-\}, \tag{8}$$

$$\{1^+0^+2^-\} \ - \ \{\,0^+2^+1^-\}. \tag{9}$$

The coefficients here do not assist us in making L a good quantum number: they are chosen purely on the basis of the permutational symmetry of the m_s values with respect to the m_ℓ values. In fact, if we use the notation of Rutherford,(9) namely

$\beta \equiv$ 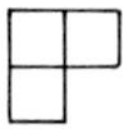 , $s \equiv \begin{smallmatrix}1&3\\2&\end{smallmatrix}$, $r \equiv \begin{smallmatrix}1&2\\3&\end{smallmatrix}$,

then it turns out that (8) and (9) correspond respectively to

$$e^{\beta}_{ss}\ \{\,2^+1^+0^-\}, \qquad e^{\beta}_{sr}\{2^+1^+0^-\},$$

where e^{β}_{ss} and e^{β}_{sr} are the semi-normal permutational operators of Young.

One advantage of Harter's method is at once clear from the example above: for $M_L = 3$ and $S = \frac{1}{2}$ there are only three tableaux ($\begin{smallmatrix}1&1\\4&\end{smallmatrix}$, $\begin{smallmatrix}1&3\\2&\end{smallmatrix}$, and $\begin{smallmatrix}1&2\\3&\end{smallmatrix}$) but four determinants. This reduction in the number of basis states for a given S becomes more marked for smaller M_L. Of course, it is part of Harter's program to avoid using Slater determinants altogether. To this end, he had developed a calculus to enable any required matrix element to be worked out directly from the states defined by the tableaux. Twenty-five years ago such techniques would not have been considered useful because of their tedious and repetitious character. Harter argues, however, that the mechanical nature of the calculations become a positive advantage when programming a computer. No cfp have to be worked out; no elaborate subroutines for calculating 6-j and 9-j symbols are needed; and all multiplicity difficulties can be settled by straightforward algorithms. Powerful though these arguments are, they can equally well be made for using Slater determinants; and the user of that classic approach can altogether avoid Harter's elaborate apparatus, with its jawbone formulae, hook-lengths, Gelfand states, and other unconventional features.

6. CONCLUSION

Methods other than the ones described here have been devised. For example, we may take Slater determinants to play the role of intrinsic states (as in nuclear theory), forming states of well defined S and L by integrating over their (weighted) rotated forms. (10) But, by and large, the techniques of Secs. 2-5 provide us with a sufficiently large choice of options — as well as an excellent source of independent checks. The azimuthal quantum number ℓ largely determines the approach. For small ℓ (0 or 1), the method of Slater determinants is adequate: Racah's methods become advantageous for $\ell = 2$ or 3: while for larger ℓ (if such calculations should ever prove worth carrying out) the quasiparticle approach comes into its own. It is probably too early to estimate the impact that Harter's method will have on atomic physics. Much depends on the willingness and adaptability of physicists and chemists to try it out.

The work reported in this paper was supported in part by the United States National Science Foundation.

REFERENCES

1. G. Racah, Phys. Rev. 62, 438 (1942); 63, 367 (1943); 76, 1352 (1949).
2. B. R. Judd, Phys. Rev. 173, 40 (1968).
3. M. Rotenberg, R. Bivins, N. Metropolis, and J. K. Wooten Jr., The 3-j and 6-j Symbols, MIT Press, Cambridge, Mass. (1959).
4. C. W. Nielson and G. F. Koster, Spectroscopic Coefficients for the p^n, d^n, and f^n Configurations, MIT Press, Cambridge, Mass. (1963).
5. C. L. B. Shudeman, J. Franklin Inst. 224, 501 (1937).
6. L. Armstrong Jr. and B. R. Judd, Proc.Roy. Soc. (London) A315, 27 (1970); A315, 39 (1970).
7. W. G. Harter, Phys. Rev. A8, 2819 (1973).
8. W. G. Harter and C. W. Patterson, A Unitary Calculus for Electronic Orbitals, Lecture Notes in Physics (Springer-Verlag) 49 (1976).
9. D. E. Rutherford, Substitutional Analysis, Edinburgh University Press (1948).
10. B. R. Judd, Zeits. f. Phys. A278, 117 (1976).

Double Cosets and the Evaluation of Matrix Elements

T.H. Seligman
Instituto de Fisica
Universidad Nacional Autónoma de México

Talk presented at the Symposion on the permutation group and it's applications in Chemistry and Physics, Bielefeld 1978.

Abstract

We discuss the use of double coset decompositions for the evaluation of quantum mechanical matrix elements between states symmetry adapted to some finite group. In order to display the essential features we concentrate on the simple case of operators that are invariant under the action of the group, and primitive states that are invariant under the action of a subgroup. By way of example the normalization matrix element for n-body states symmetry adapted to an IR of S_n are calculated.

I. Introduction.

In this lecture we shall present, in its simplest form, a method to calculate matrix elements between states, symmetry adapted to some finite group in general, and to the permutation group S_n in particular.

The realm of applications contains two important fields: The molecular point group symmetry and the permutational symmetry related to the Pauli principle. The general method applies to both cases equally well, but as far as examples are concerned, we shall concentrate on the latter, where the practical advantage may be larger.

We shall not stop here to discuss the numerous alternative methods to solve the problems discussed, but instead limit ourselves to note that the method presented here is completely general. It may not always be the most practical one for a given problem, particularly if it may readily be treated by "pedestrian" methods.

The method to be presented is based on double coset decompositions of the group involved, and it's development was initiated independently for electron (1,2,3) and nucleon systems (4,5). Problems involving the point group symmetry of molecules were discussed more recently (6,7). An extensive review article including this subject, as well other applications of double coset decompositions of finite groups is forthcoming (7).

The basic idea involved is simple: assume we wish be calculate a matrix element between states, transforming according to a given IR of a group G, of an operator that is invariant under the action of G. Assume further that we obtain these states by applying a projection operator to what we may call a basis of "primitive" states invariant under some subgroup $H \subset G$. We then have to evaluate a matrix element of the projector and the invariant operator (that commutes with the projector) between primitive states. Clearly any element of H can be trivially applied both to bra and ket, and we thus only have to consider the double coset (DC) generators of the decomposition $H \setminus G / H$ of G with respect to H on both sides.

The general method as described in (7) is valid under much weaker assumptions. The primitive states in bra and ket may belong to any IR of two different groups $H, K \subset G$, and the operator may be invariant under a subgroup of G rather then under G, or it may even be a tensor operator for such a subgroup. If we consider such general cases we obtain a generalized Wigner-Eckhardt theorem (7).

The purpose of this lecture is not to give the general theory, but rather to indicate, using the simple example sketched above, how the basic ideas involved, may be understood and applied. Furthermore the case, where the primitive functions are invariant with respect two the subgroup H, is practically the most important one. We shall only release the restriction that the invariance group of the primitive states in bra and ket be equal.

In the next section we define the problem properly, in order to obtain the main result. In section IV we specialize to the case of S_n and discuss matrix elements between n-particle states of given orbital symmetry.

II. Symmetry adapted states and their matrix elements.

We shall work in a Hilbert space whose elements we denote by kets $| \, >$. In this space a group G of (unitary) operators g acts. We shall denote its IR by α with elements $\Gamma^{\alpha}(g)$, $g \in G$ and by d_{α} the dimension of the IR α.

Assume that for some physical reason we are interested in states that transform according two the unitary IR α of G i.e.

$$g|\gamma;\alpha r> = \sum_{s=1}^{d_{\alpha}} |\gamma;\alpha s> \Gamma^{\alpha}_{s,r}(g), \tag{2.1}$$

where s, r are row and column labels of $\Gamma^{\alpha}(g)$ and characterizes all properties of the state not defined by other labels.

To obtain such states from an arbitrary state the standard method is to apply matric basis elements defined by

$$e_{r,s} = \frac{d_{\alpha}}{|G|} \sum_{g \in G} \Gamma^{\alpha}_{s,r}(g^{-1})\, g, \tag{2.2}$$

where $|G|$ denotes the order of the group G. It may be easily checked (3,5,7) from the representation property of $\Gamma^{\alpha}(g)$ that they fulfill the relations

$$e^{\alpha}_{r,s}\, e^{\beta}_{t,u} = \delta_{\alpha,\beta}\, \delta_{s,t}\, e^{\alpha}_{r,u} \tag{2.3a}$$

$$g\, e^{\alpha}_{r,s} = \sum_{t} \Gamma_{t,r}(g)\, e^{\alpha}_{t,s} \tag{2.3b}$$

$$(e^{\alpha}_{r,s})^{+} = e^{\alpha}_{s,r} \,. \tag{2.3c}$$

From (2.3b) we may conclude that

$$|\gamma s;\alpha r> = e^{\alpha}_{r,s}|\gamma> \tag{2.4}$$

is either zero or transforms according to (2.1).

The role of the label s is somewhat peculiar: If the state (2.4) vanishes identically for all s, $|\gamma>$ does not contain the IR α. Otherwise the resulting states for different s may or may not be independent. If the original space has no symmetry property, they may all be independent. Note also that symmetry adapted states behave under the application of a matric basis element as

$$e^{\alpha}_{r,s}|\gamma \; ; \alpha's'> \;=\; \delta_{\alpha\alpha'} \;\; \delta_{ss'} \;|\gamma \;, \alpha \; r >. \qquad (2.5)$$

This relation sheds some light into the role of the label s and will be important later.

As we mentioned earlier we wish to use "primitive" functions that are already symmetry adapted to an IR of a subgroup $H \subset G$. This leads as to the necessity of talking about sequence adaption of states.

In a state $|\gamma \; ; \alpha r >$ we may choose the row label such that it contains the IR η of H as a label, i.e.

$$|\gamma \; ; \alpha r> \equiv |\gamma \; ; \alpha \; \sigma \; \eta \; j> \qquad (2.6)$$

where σ is a multiplicity label taking care of possible multiplicities in the subduction from α to η, and j is a row label of η. The state (2.6) is called "sequence adapted".

Such sequence adapted labels may be used in the matric basis and the representation. We then have

$$e^{\alpha}_{\sigma\eta j,\sigma'\eta'j'} \;=\; \frac{d\alpha}{|G|} \sum_{g \in G} \Gamma^{\alpha}_{\sigma'\eta'j',\sigma\eta j}(g^{-1}) \;\; g. \qquad (2.7)$$

Using matric basis elements for the subgroup also it follows that

$$e^{\alpha}_{\sigma\eta j,\sigma'\eta'j'} \; e^{\eta''}_{j''',j''} \;=\; \delta_{\eta'\eta''} \delta_{j'j'''} \; e^{\alpha}_{\sigma\eta j,\sigma'\eta''j''} \; . \qquad (2.8)$$

This together with (2.5) in turn indicates how the label s must be chosen if the primitive state is symmetry adapted to a subgroup because

$$e^{\alpha}_{\sigma\eta j,\, \sigma'\eta' j'} \; |\gamma \; ; \; \eta'' j'' >$$

$$= \delta_{\eta'\eta''}\delta_{j'j''} \; e^{\alpha}_{\sigma\eta j,\, \sigma'\eta'' j''} \; |\gamma \; ; \; \eta'' j'' > \tag{2.9}$$

$$= \delta_{\eta'\eta''}\delta_{j'j''} \; |\gamma\sigma'\eta'' j'' \; ; \; \alpha\sigma\eta j >$$

We are only free to choose σ' while η' and j' are fixed if we wish to avoid the trivial result. On the other hand states with different σ' are now independent if no additional particular property of the primitive functions come to bear; e.g. further dependences would occur if the primitive state would transform according to an IR of a group larger than H.

The concepts developed up to here require one important generalization. Actually we may want to have a different chain of groups characterizing our row label r in a state $|\gamma;\alpha \; r>$, than the one we need, to pick out the non zero contributions from the primitive functions. Starting from some standard orthonormal basis, we denote by U and V the unitary transformations that take us to the desired orthonormal bases, e.g. the sequence adapted basis of type (2.6). We then define a skew representation matrix (3) by

$$\mathbb{\Gamma}^{\alpha}(g) = U \, \Gamma^{\alpha}(g) \, V^{+} \tag{2.10}$$

and a skew matric basis element by

$$\mathbb{e}^{\alpha}_{r,s} = \frac{d_{\alpha}}{|G|} \sum_{g \in G} \mathbb{\Gamma}^{\alpha}_{s,r}(g^{-1}) \, g \tag{2.11}$$

Clearly these definitions depend on U and V. Rather than spelling this dependence out explicitely, we shall use the convention that similar labels possibly with primes refer to the same basis.

It is easy to verify that

$$\mathbb{e}^{\alpha}_{r,s}\mathbb{e}^{\alpha'}_{s',t} = \mathbb{e}^{\alpha}_{r,t} \, \delta_{ss'} \, \delta_{\alpha\alpha'} \tag{2.12a}$$

holds. If s and s' had not been selected from the same basis an overlap $\mathbb{\Gamma}^{\alpha}_{ss'}(e)$ (e being the identity in G) would have appeared in lieu of the Kronecker δ. The above relation generalizes (2.3a). (2.3b,c) generalize similarly yielding

$$g\,\mathbb{E}^{\alpha}_{r,s} = \sum_{r'} \Gamma^{\alpha}_{r'r}(g)\,\mathbb{E}^{\alpha}_{r',s} \tag{2.12b}$$

$$(\mathbb{E}^{\alpha}_{r,s})^{+} = \mathbb{E}^{\alpha}_{s,r}\ . \tag{2.12c}$$

Note that in (2.12b) a standard and not a skew representation matrix appears. We can thus define a state symmetry adapted from some primitive state as

$$|\gamma\sigma\eta j\ ;\ \alpha r> \;=\; \mathbb{E}^{\alpha}_{r,\sigma\eta j}|\gamma\ ;\ \eta j\ >. \tag{2.13}$$

We are now in a position to write down matrix elements and, as mentioned earlier, we shall restrict ourselves to the particular case of an operator that is invariant under the group with respect to which we symmetry adapt. Furthermore we assume that the primitive states are invariant under groups H or K; i.e. their IR η and μ are one-dimensional. We can then drop the corresponding row labels. Thus we have

$$<\bar{\gamma}\ \rho\ \mu\ ;\ \alpha\ r|\mathcal{O}|\ \gamma\ \sigma\ \eta\ ;\ \alpha' r'\ >$$

$$= <\bar{\gamma}\ ;\mu|(\mathbb{E}^{\alpha}_{r,\rho\mu})^{+}\mathcal{O}\,\mathbb{E}^{\alpha'}_{r',\sigma\eta}|\gamma\ ;\ \eta\ > \tag{2.14}$$

$$= <\bar{\gamma}\ ;\mu|\ \mathcal{O}\,\mathbb{E}^{\alpha}_{\rho\mu,\sigma\eta}\ |\gamma\ ;\ \eta\ >\ \delta_{\alpha\alpha'}\delta_{rr'}$$

where in the first step we used the definition (2.13) and in the next step relation (2.12b,c) as well as the invariance of $\mathcal{O}$. Now we note that $\mathbb{E}^{\alpha}_{\rho\mu,\sigma\eta}$ is a linear combination of group elements of G. On the other hand we have $h|\gamma\ ;\eta> \;=\; |\gamma;\eta>$ if $h \in H \subset G$ and similarly for elements of K on $|\bar{\gamma};\mu>$. It is therefore convenient to perform a double coset (DC) decomposition of G as

$$G \;=\; Kg_1H \;\oplus\; Kg_2H \;\oplus\; \ldots\; Kg_mH, \tag{2.15}$$

where m is the number of disjoint DC and g_i are the DC generators.

In distinction to the more common right and left cosets each element in Kg_iH may be covered several times if in kg_ih , h and k cover independently all elements of H and K. The number of times an element occurs is constant within each DC and is given by

$$M(i) = |K \cap g_i H g_i^{-1}| \tag{2.16}$$

The choice of the generator g_i for the ith DC is arbitrary, as it can be any element in this DC, but the results we obtain do not depend on this choice.

We now write the skew matric basis element appearing in (2.14) explicitely as

$$\mathcal{C}^{\alpha}_{\rho\mu,\sigma\eta} = \frac{d\alpha}{|G|} \sum_{g \in G} \Gamma^{\alpha}_{\sigma\eta,\rho\mu}(g^{-1})\, g \tag{2.17}$$

$$= \frac{d\alpha}{|G|} \sum_{i=1}^{m} \frac{1}{M(i)} \sum_{k \in K,\, h \in H} \Gamma^{\alpha}_{\alpha\eta,\rho\mu}(h^{-1} g_i^{-1} k^{-1})\, k g_i h$$

$$= \frac{|H|\,|K|\,d\alpha}{|G|} \sum_{i=1}^{m} \frac{1}{M(i)} \Gamma^{\alpha}_{\sigma\eta,\rho\mu}(g_i^{-1})\, P^{\mu} g_i P^{\eta}$$

where

$$P^{\mu} = \frac{1}{|K|} \sum_{k \in K} k, \quad P^{\eta} = \frac{1}{|H|} \sum_{h \in H} h \tag{2.18}$$

are the projectors on the identical IR μ and η of K and H respectively. These in turn are absorbed by the primitive states in bra and ket and we obtain

$$\langle \bar{\gamma}\rho\mu;\, \alpha r | \mathcal{O} | \gamma\sigma\eta\,;\alpha' r' \rangle$$

$$= \frac{|H|\,|K|\,d_{\alpha}}{|G|} \sum_{i=1}^{m} \frac{1}{M(i)} \Gamma_{\sigma\eta,\rho\mu}(g_i^{-1}) \tag{2.19}$$

$$\cdot \langle \bar{\gamma};\, \mu | \mathcal{O} g_i | \gamma;\, \eta \rangle \quad \delta_{rr'} \delta_{\alpha\alpha'}$$

The final result is not only diagonal in r but also independent of this label, as we would expect from Wigners theorem. Also the coefficient $\Gamma_{\sigma\eta,\rho\mu}(g_i^{-1})$ only depend on the DC and not on the representative g_i . The same holds for the remaining matrix elements between primitive states.

The most important point about this result is that the minimal number of independent matrix elements between primitive states appears on the right hand side, if we are not given any further information about the states or the operator. We thus have obtained the greatest simplification possible by algebraic means. The question, how to calculate the remaining coefficients and the matrix elements between primitive states, depends on the specific group and the functions used. An important example will be considered in the next section.

To conclude this discussion, we just wish to repeat that the result may be generalized - at the expense of writing many labels - to more complicated operators and primitive states (7).

III. Matrix elements between states of well defined permutational symmetry.

A case of particular importance arises from the Pauli principle. Indeed the latter requires taking matrix elements between antisymmetric states, or more generally, if we can separate the dependence on some intrinsic variables (spin for electron, spin and Iso-spin for nucleons), between states transforming according to some IR of S_n.

Due to these intrinsic variables, we can have primitive orbital functions with up to two or four particles (respectively for electrons or nucleons) in a state symmetric with respect to the interchange of orbital variables. In a single particle picture this simply reduces to the possible multiple occupancy of orbitals.

Although it is not essential to most of our argument, we shall proceed in a single particle basis, as this case is both very illustrative and of great importance. We shall though allow this basis to be non-orthogonal.

A primitive orbital function - that is function for which we can perform integrals readily - will obviously be chosen as a product of single particle orbitals.

We shall designate by n_i the occupation number of the ith orbital Ψ_i and by $\underline{n}_i$ the set of particle labels of the particles in this orbital.

A primitive orbital function may be written as

$$\Psi_1\,(X_1)\ \Psi_1\,(X_2)\ \cdots\ \Psi_1(X_{n_1})\ \Psi_2\,(X_{n_1+1})\ \cdots$$
$$\cdots\ \Psi_2(X_{n_1+n_2})\ \cdots\ \Psi\ (X_{n-n\ \ +1})\ \cdots\ \Psi\ (X_n) \tag{3.1}$$

if we have $\varkappa$ orbitals and $n = \sum_{i=1}^{\varkappa} n_i$ particles. This function is clearly invariant under the group

$$H = S_{n_1} \times S_{n_2} \times \cdots S_{n_\varkappa} \ . \tag{3.2}$$

Here S_{n_i} indicates the group of permutations of the labels in the set $\underline{n}_i$. H clearly is a direct product of such groups.

Invariance with respect to H implies invariance with respect to each factor and we shall write the primitive states of (3.1) in short as

$$|\gamma;n> \;\equiv\; |\gamma\;;\;\{n_1\}\;\{n_2\}\cdots\{n_{\chi}\}> \tag{3.3}$$

where $\{n_i\}$ stands as usual for the identical or symmetric representation of S_{n_i}. γ contains all the information about the single particle orbitals, as we will not need this information to follow the argument of the previous section. We can then write the symmetry adapted state

$$|\gamma\sigma\eta\;;\;\alpha\,r> \;=\; \mathfrak{C}^{\alpha}_{r,\sigma\eta}|\gamma,\eta> \tag{3.4}$$

where $\eta = \{n_1\}\{n_2\}\cdots\{n_{\chi}\}$ is the identity IR of H while α is some IR of S_n. r can be any type of row label and α any multiplicity label.

Next we introduce a similar set of definitions for the state in the bra of the matrix element using a (possibly different) simple particle basis $\bar\Psi_i$; i=1 $\cdots$ χ and a set of different occupation numbers $\bar n_1\;\bar n_2\cdots\bar n_{\chi}$. This leads to primitive states

$$|\bar\gamma\;;\;\mu> \;=\; |\bar\gamma\;;\{\bar n_1\}\;\{\bar n_2\}\cdots\{\bar n_{\chi}\}> \tag{3.5}$$

$$=\;\bar\Psi_1(X_1)\cdots\bar\Psi_1(X_{\bar n_1})\;\bar\Psi_2(X_{\bar n_1+1})\cdots\bar\Psi_{\bar\chi}(X_n)$$

with an invariance group

$$K \;=\; S_{\bar n_1}\times S_{\bar n_2}\times\cdots S_{\bar n_{\bar\chi}}\;. \tag{3.6}$$

Using (2.19) we get for a matrix element of a symmetric operator between symmetry adapted states

$$<\bar\gamma\rho\mu\;;\alpha\,r|\mathcal{O}|\;\gamma\sigma\eta\;;\alpha'r'> \;=\; \delta_{\alpha\alpha'}\;\delta_{rr'} \tag{3.7}$$

$$\frac{\prod_{i=1}^{\chi} n_i!\;\prod_{j=1}^{\bar\chi}\bar n_j!}{n!}\;d_{\alpha}\sum_{i=1}^{m}\frac{1}{M(i)}\;\Gamma_{\sigma\eta,\rho\mu}(g_i^{-1})\;<\bar\gamma;\mu|\mathcal{O}g_i|\gamma;\eta>$$

where the g_i are DC generators for the decomposition $K\backslash S_r/H$. Clearly we have separated the algebraic part contained in the factors $1/M(i)$ and $\Gamma_{\sigma\eta,\rho\mu}(g_i^{-1})$ from the integrals $<\gamma;\mu|\mathcal{O}g_i|\gamma;\eta>$.

Before we can proceed any further we have to characterize the DC we use. This fortunately is possible using a result of Coleman (8):
We define a $\chi \times \bar{\chi}$ matrix D called "DC symbol" as a matrix of positive integers D_{ij} that fulfill

$$\begin{aligned} \sum_i D_{ij} &= n_j \\ \sum_j D_{ij} &= \bar{n}_i \end{aligned} \tag{3.8}$$

It may be shown that the DC symbols are in one to one correspondence with the DC $K \backslash S_n / H$ if K and H are defined as in (3.6) and (3.2). Furthermore (9,5) we can give a set of DC generators by segmenting the sets $\underline{n}_j$ into subsets $\underline{n}_{ij}$ of length D_{ij} and defining

$$g_D = \begin{pmatrix} \underline{n}_{11}\ \underline{n}_{21} \cdots \underline{n}_{\bar{\chi}1}\ \underline{n}_{12} \cdots\cdots \underline{n}_{1\chi} \cdots \underline{n}_{\bar{\chi}\chi} \\ \underline{n}_{11}\ \underline{n}_{12} \cdots \underline{n}_{1\chi}\ \underline{n}_{21} \cdots\cdots \underline{n}_{\bar{\chi}1} \cdots \underline{n}_{\bar{\chi}\chi} \end{pmatrix} \tag{3.9}$$

The symbol on the right hand side has to be read as the usual two row symbol for permutation, after each set $\underline{n}_{ij}$ has been replaced by the proper string of numbers. In the upper row the numbers $12\cdots n$ will appear in natural order, while below some permutation occurs. As each DC is characterized by a DC symbol D, we label the DC generator by this symbol.

We are now in a position to evaluate the integral occurring in (3.8) if we choose a given operator $\mathcal{O}$. As an example we choose the identity operator, that is particularly simple and yields the important normalization matrix elements.

The DC generator as defined in (3,9) clearly transfers D_{ij} particles that occupied the orbital Ψ_j to the orbital $\bar{\Psi}_i$. Upon taking the integral over the variables with labels in $\underline{n}_{ij}$ this clearly yields the overlap $<\bar{\Psi}_i|\Psi_j>$ for each of these variables. As we have product wave functions this overlap occurs to the power D_{ij} , and we thus obtain

$$\langle \bar{\gamma};\mu|g_D|\gamma\ ;\ \eta \rangle = \prod_{i=1}^{\bar{\varkappa}} \prod_{j=1}^{\varkappa} \langle \bar{\Psi}_i | \Psi_j \rangle^{D_{ij}} \tag{3.10}$$

A similar argument shows us that the coefficient $\mathbf{M}(D)$ (also only dependent on the DC and thus labeled by D) is given by

$$M(D) = \prod_{i=1}^{\bar{\varkappa}} \prod_{j=1}^{\varkappa} D_{ij}! \tag{3.11}$$

The evaluation of the representation matrix elements $\Gamma^{\alpha}_{\sigma\eta,\rho\mu}(g_i^{-1})$ clearly still depends on the choice of the multiplicity labels. A particularly useful choice can be made, if we consider that the chain of groups

$$S_n \supset S_{n-n} \times S_n \supset \cdots \supset S_{n_1+n_2} \times S_{n_3} \times \cdots S_n \subset H \tag{3.12}$$

affords a multiplicity free subduction to the identity IR of H, which in turn implies the identity IR $\{n_i\}$ for each S_{n_i}. This results from the fact, that at each step the subduction breaks an arbitrary IR of $S_{n_1+n_2+\cdots n_i}$ into an arbitrary IR of $S_{n_1+n_2+\cdots n_{i-1}}$ and a completely symmetric IR of S_{n_i}; but this subduction is known to multiplicity free (10).

The string

$$\sigma \equiv \alpha_2,\alpha_3 \cdots \alpha_{\varkappa-1} \tag{3.13}$$

of IR α_i of $S_{n_1+n_2\cdots n_i}$ may thus be chosen for the multiplicity labels σ and similarly for ρ.

With this choice closed formulae (<u>11</u>) and generating functions (<u>5</u>.<u>11</u>) for the representation coefficients are given in the literature. Thus both the integrals and the coefficients appearing in the central result (3.7) are readily available.

Note that we made explicit use of the single particle basis <u>only</u> in eq. (3.10). Eq. (3.7) thus holds true for <u>any</u> set of primitive functions that are invariant under H and K respectively. The question how to evaluate the integral then has to be reexamined. One such case is encountered in the nuclear cluster model, where translationally invariant functions are used. Also in this case the DC symbol plays a central role in finding the integrals (12).

If we have an operator different from unity the integral is more complicated. Yet if it is e.g. a sum of two-body operators, techniques similar to the ones above can be used for further reduction (4,5). In the case of single particle orbitals the problem is indeed reduced to - in the worst case - four-center integrals for the interaction and overlap integrals.

IV. Conclusion

The double coset decomposition seems to be the natural framework to simplify quantum mechanical matrix elements related in any way with a group of operators. This stems from the very structure of a matrix element; namely from the fact that the operator is sandwiched between two functions, that may be invariant or covariant with respect to some group. In this sense the use of double cosets is not only a useful trick, but turns out to be implicit in the very nature of a quantum mechanical calculation.

This becomes apparent if we view the example discussed in section III in a somewhat wider scope: if we have orthonormal single particle functions the normalization problem becomes trivial and even two-body matrix elements are easily obtained by fractional parentage techniques. If we proceed to the next degree of complication, namely non-orthogonal orbitals, there are still powerful methods, involving Slater determinants or second quantization, available. Once we leave the single particle picture, only the method developed here remains adequate. This method is thus the only one exclusively based on the symmetries of the problem, without making use of other particular properties of the orbital function.

I wish to thank D. Klein, with whom I have discussed many hours on double cosets, and on whose work much of the material presented is based.

Literature

1. C. Herring, Rev. Mod. Phys. 34 (1962) 631.
2. B.R. Junker and D.J. Klein, J. Chem. Phys. 55 (1971) 5533; F.A. Matsen, D.J. Klein and D.C. Foyt, J. Chem. Phys. 75 (1971) 1866; D.J. Klein and W.A. Seitz, Phys. Rev. B8 2276 (1973).
3. D.J. Klein, "Group Theory and its Application III; ed. E.M. Loebel, Academic Press, New York, 1975.
4. P. Kramer and T.H. Seligman, Nucl. Phys. A136 (1969) 545; Nucl. Phys. A186 (1972) 49.
5. T.H. Seligman, "Double Coset Decompositions of Finite Groups and the Many-Body Problem" Burg Monographs in Science, Vol. I; ed. D. Clement, Burgverlag, Basel, 1975.
6. E.R. Davidson, J. Chem. Phys. 62 (1975) 400.
7. W. Hässelbarth, D.J. Klein, E. Ruch and T.H. Seligman, (in preparation).
8. J. Coleman, Queens Lectures (1956).
9. P. Kramer, Z. Physik 216 (1968) 68.
10. M. Hammermesh, "Group Theory" Academic Press, New York, 1959.
11. A. Antillon, Thesis, Mexico 1978; A. Antillon, G. Lopez and T.H. Seligman (in preparation).
12. T.H. Seligman and W. Zahn, J. Phys. G2 (1976) 79; H.H. Hackenbroich, T.H. Seligman and W. Zahn, Helv. Phys. Acta 50 (1977) 723.

Properties of Double Cosets with Applications to Theoretical Chemistry

J. S. Frame
Michigan State University

1. Double cosets and structures.

A useful model of a molecule is a collection of N atoms or groups of atoms called ligands, each oscillating with small amplitude about an equilibrium position or site, where the N sites form a rigid geometrical figure called a skeleton, that may itself be subject to uniform translation and rotation in space (3,6). An important classification problem is the determination of the number of non-equivalent configurations having the same skeleton and the same numbers of ligands of each kind.

If the sets of N sites and N ligands are each arbitrarily assigned labels from 1 to N, a configuration is determined by a permutation x that maps each site number onto the ligand number at that site. We may write

$$x = \begin{pmatrix} 1 & 2 & \cdots N \\ 1_x & 2_x & \cdots N_x \end{pmatrix} \tag{1.1}$$

where k_x denotes the ligand number at site k, and the columns in the symbol for x can be arranged arbitrarily. Each permutation x, of the group $G = S_N$ of all N permutations on N symbols, describes a configuration. If the skeleton has symmetry, each symmetry operation is described by a permutation $\underline{a}$ of the site labels and these permutations form a subgroup A of G. Similarly, the permutations $\underline{b}$ of ligand labels that rearrange indistinguishable ligands form a subgroup B of G. All configurations indistinguishable from the one labeled x belong to the double coset $A \times B$, consisting of all elements $a \times b$ of G with $a \in A$ and $b \in A$ (6). These will be studied further in §3, where we obtain counting formulas for the number of configurations, and for the number of modes with distinct ligands.

First we consider in §2 the group matrices (4)

$$\bar{A} = \sum \alpha_i^{-1} a_i \ , \ \bar{B} = \sum \beta_j^{-1} b_j \tag{1.2}$$

of linear representations $a_i \to \alpha_i$ and $b_i \to \beta_j$ of the subgroups A and B mapping $a_i \in A$ and $b_j \in B$ onto roots of unity α_i, β_j. Then we study the group matrices $\bar{M}$ and $\bar{N}$ of monomial representations of G of dimensions m and n, namely $x \to M_x$ and $x \to N_x$, induced by $\bar{A}$ and $\bar{B}$, for $x \in G$. The m x n intertwining matrices V such that

$$\bar{M}V = V\bar{N} \tag{1.3}$$

form a vector space $\mathcal{V}$ having a basis V^h, one for each non-vanishing weighted double coset $\bar{A}g_h\bar{B}$. The dimension d of $\mathcal{V}$ reduces to the number of ordinary double cosets when $\alpha_i = \beta_j = 1$ for all i,j. This number can be computed from the characters of the induced representations, and be expressed in terms of the distribution of double coset elements in conjugacy classes C_k of G.

In §4 we consider a second application of double cosets to molecular structures, related to the calculation of normal coordinates that yield a diagonalized potential energy matrix, and the selection of a correct skeleton from possible alternatives by observing the molecular spectrum. The idea, which will be explored more fully, is to choose 3N appropriately weighted displacement doordinates x_i of ligands from corresponding sites, so that the kinetic energy E_K and potential energy E_V are approximated by the quadratic forms

$$E_K = \dot{X}^T\dot{X}/2, \quad E_V \doteq X^TVX/2 \tag{1.4}$$

where $\dot{X}$ denotes dX/dt and X is a 3N column vector of x_i. If a matrix T could be found such that

$$T^TT = I, \quad T^TVT = \Lambda = \text{diag}\,\{\lambda_i\} \tag{1.5}$$

then the transformation X = TQ to "normal coordinates" q_i would simplify the equations of motion to

$$d^2q_i/dt^2 + \lambda_i q_i = 0. \tag{1.6}$$

The eigenvalues λ_i could then be obtained from the frequencies $\lambda_i^{1/2}/2\pi$

observed in molecular spectra. Criteria are developed to accept one possible skeleton and reject another. But the required matrix T, whose columns are eigenvectors of the unknown matrix V, can be found from certain double coset matrices associated with the symmetry group G of the molecule, and V can be constructed from T and Λ.

In §5 we illustrate the explicit computation of T for the case of a methane molecule with tetrahedral skeleton.

2. Group matrices and double coset matrices.

Let G be a finite group of order $^\circ G$, and F a complex field containing all the $^\circ G$th roots of unity. The group ring FG of G over F consists of all linear combinations $\sum c_i g_i$ for $g_i \in G$ and $c_i \in F$, where c_i commutes with each g_h. The projection operator E is here defined to be a mapping from FG to F that picks out from each ring element the coefficient of the identity element g_1:

$$E(\sum c_i g_i) = c_1. \tag{2.1}$$

The group matrices $\bar{A} = \sum \alpha_i^{-1} a_i$ and $\bar{B} = \sum \beta_j^{-1} b_j$ are ring elements, and for $g_h \in G$ the ring element $\bar{A} g_h \bar{B}$ is called a weighted double coset (1,4). If the intersection $D_h = A \cap g_h B g_h^{-1}$ of subgroups A and $g_h B g_h^{-1}$ has order $^\circ D_h$, then the coefficient of g_h in $\bar{A} g_h \bar{B}$ is the coefficient $E[\bar{A} g_h \bar{B} g_h^{-1}]$ of g_1 in $\bar{A} g_h \bar{B} g_h^{-1}$. Only those elements of $g_h B g_h^{-1}$ in D_h have inverses in A. If $\bar{D}_h$ is the restriction to D_h of $g_h \bar{B} g_h^{-1}$

$$E(\bar{A} g_h \bar{B} g_h^{-1}) = E(\bar{A}\bar{D}_h). \tag{2.2}$$

This is $^\circ D_h$ or 0 according as the restriction of $\bar{A}$ to D_h is $\bar{D}_h$ or not. Thus, either $\bar{A} g_h \bar{B}$ is 0, or every group element in $\bar{A} g_h \bar{B}$ has a coefficient of absolute value on $^\circ D_h$. We define

$$K_h = \bar{A} g_h \bar{B} / {}^\circ D_h \tag{2.3}$$

to be a proper double coset if it is not 0, and write

$$K_h' = \bar{B} g_h^{-1} \bar{A} / {}^\circ D_h \tag{2.4}$$

for the related linear combination of the inverse elements. Let the subgroup A of order $^\circ A$ have index $^\circ G / {}^\circ A = G{:}A = m$ in G, let B of order $^\circ B$ have index $^\circ G / {}^\circ B = G{:}B = n$ in G, and let $r_1 = 1$, $r_2 \ldots r_m$ and $s_1 = 1$, $s_2 \ldots s_m$ be coset representatives of A and B in G so that

$$G = A \dot{+} Ar_2 \dot{+} \ldots \dot{+} Ar_m = B \dot{+} Bs_2 \dot{+} \ldots \dot{+} Bs_n \tag{2.5}$$

Then the $m \times n$ matrix V^h with (i,j)-entry

$$V^h_{ij} = E(s_j^{-1}K'_k r_i) \tag{2.6}$$

which is the coefficient in K'_h of $(r_i s_j^{-1})^{-1}$ is called the double coset matrix belonging to K_h (2,4).

The group matrix $\bar{M}$ of a representation $x \longrightarrow M_x$ of G by $m \times m$ matrices M_x is defined by

$$\bar{M} = \sum_{x \in G} x^{-1}M_x = \sum_{x \in G} x\, M_x^{-1} . \tag{2.7}$$

It is a matrix with entries from FG having the properties

$$g\bar{M} = \bar{M}M_g \;,\; \bar{M}g = M_g\bar{M} \tag{2.8}$$

$$E(g\bar{M}) = E(\bar{M}g) = M_g \tag{2.9}$$

The $m \times m$ monomial matrices M_g of the representation $g \longrightarrow M_g$ of G induced from the linear representation $a_i \longrightarrow \alpha_i$ of the subgroup A of index m has (i,j)-entry

$$(M_g)_{ij} = E(\bar{A}_j r_i g r_j^{-1}) = E(r_j^{-1}\bar{A} r_i g) \tag{2.10}$$

This is α_k if $r_i g = a_k r_j$ with $a_k \in A$, and 0 otherwise, so there is just one non-zero entry in each row and column. It is easily shown that

$$M_x M_y = M_{xy} \text{ for } x,y \in G. \tag{2.11}$$

Similarly, the $n \times n$ matrix N_g in the monomial representation of G induced from $\bar{B}$ has the (i,j)-entry

$$(N_g)_{ij} = E(\bar{B}s_i g s_j^{-1}) = E(s_j^{-1}\bar{B}s_i g) \tag{2.12}$$

and the group matrix is

$$\bar{N} = \sum_{x \in G} x^{-1}N_x = \sum_{x \in G} xN_x^{-1}. \tag{2.13}$$

Definition 2.1. The $m \times n$ matrix V (over F) intertwines the group matrices $\bar{M}$ and $\bar{N}$ if and only if

$$\bar{M}V = V\bar{N} . \tag{2.14}$$

Since linear combinations of intertwining matrices with coefficients from F also intertwine, these matrices V form a vector space $\mathcal{V}$ over F.

Theorem 2.1. The proper double coset matrices V^h defined in (2.6) are a basis for the vector space $\mathcal{V}$. Each matrix $V = (v_{k\ell})$ of $\mathcal{V}$ has the form

$$V = \sum_h c_h V^h , \quad c_h = \sum_{k\ell} (v_{k\ell}/{}^\circ G) E(r_k^{-1}\bar{A}g_h\bar{B}s_\ell). \tag{2.15}$$

The dimension d of $\mathcal{V}$ is the number of independent double cosets $\bar{A}g_h\bar{B}$.

Proof: Equation (2.14) implies

$$M_x V = VN_x \quad \text{or} \quad V = M_x^{-1} V N_x \quad \text{for all} \quad x \in G. \tag{2.16}$$

Writing $V = (V_{k\ell})$ we sum over x and divide by ${}^\circ G$.

$$V = (1/{}^\circ G) \sum_{x \in G} M_x^{-1} V N_x = \sum_{M,\ell} (v_{k\ell}/{}^\circ G) V^{k\ell} \tag{2.17}$$

where the $m \times n$ matrix $V^{k\ell}$ has the (i,j)-entry

$$\begin{aligned} v_{ij}^{k\ell} &= \sum_{x \in G} (N_x)_{\ell j} (M_x^{-1})_{ik} \\ &= \sum_{x \in G} E(s_j^{-1}\bar{B}s_\ell x) E(r_k^{-1}\bar{A}r_i x^{-1}) \\ &= E(s_j^{-1}\bar{B}s_\ell r_k^{-1}\bar{A}r_i). \end{aligned} \tag{2.18}$$

If the group element $r_k s_\ell^{-1}$ has the factorization

$$r_k s_\ell^{-1} = a_p^{-1} g_h b_q, \quad a_p \in A, \ b_q \in B \tag{2.19}$$

in the double coset Ag_hB to which it belongs, then

$$V_{ij}^{k\ell} = E(s_j^{-1}\bar{B}b_q^{-1}g_h^{-1}a_p\bar{A}r_i) = \alpha_p\beta_q^{-1}\circ D_h V_{ij}^h \tag{2.20}$$

where V_{ij}^h is defined in (2.6) and

$$\alpha_p\beta_q^{-1}\circ D_h = E(r_k^{-1}\bar{A}g_h\bar{B}s_\ell) \quad \text{if} \quad \bar{A}g_h\bar{B} \neq 0. \tag{2.21}$$

Formula (2.15) follows from (2.17), (2.20) and (2.21). Since the non-vanishing matrices V^h are linearly independent and form a basis for $\mathcal{V}$; their number is the dimension d of $\mathcal{V}$.

A second formula for the dimension d of $\mathcal{V}$ follows.

Theorem 2.2. If the irreducible group matrices $\bar{\Gamma}_r$ of G occur with multiplicities μ_r^M and μ_r^N as constituents of $\bar{M}$ and $\bar{N}$, then the dimension of the intertwining space $\mathcal{V}$ is

$$d = \sum_r \mu_r^M \mu_r^N \tag{2.22}$$

Proof: Choosing matrices P and Q that completely reduce $\bar{M}$ and $\bar{N}$ into direct sums of irreducible group matrices $\bar{\Gamma}_r$ of dimension f_k where similar components are identical, we have

$$(P^{-1}\bar{M}P)P^{-1}VQ = (P^{-1}VQ)Q^{-1}\bar{N}Q. \tag{2.23}$$

so $P^{-1}QV$ intertwines $P^{-1}\bar{M}P$ and $Q^{-1}\bar{N}Q$. Now by Schur's lemma a matrix that intertwines two irreducible group matrices $\bar{\Gamma}_r$ and $\bar{\Gamma}_s$ is 0 unless they are similar, and is a scalar multiple $\lambda_r I$ of I if they are identical. Hence,

$$P^{-1}VQ = \cdot \sum I_{f_r} \times \Lambda_r \tag{2.24}$$

where Λ_r is a μ_r^M by μ_r^N matrix over F. Since these matrices (2.24) and only these intertwine $P^{-1}\bar{M}P$ and $Q^{-1}\bar{N}Q$, the dimension of $\mathcal{V}$ is given by (2.22). Every $m \times n$ matrix V of rank $r > 0$ has a factorization

$$V = PDQ^{-1} \tag{2.25}$$

where P and Q are orthogonal and D has vanishing entries except for its first r diagonal entries. In terms of these matrices P,Q equation (2.23) becomes

$$P^{-1}\bar{M}P \cdot D = DQ^{-1}\bar{N}Q. \tag{2.26}$$

We partition off the first r rows and columns in (2.26) and write

$$\begin{bmatrix} \bar{M}_{11} & \bar{M}_{12} \\ M_{21} & \bar{M}_{12} \end{bmatrix} \begin{bmatrix} D_{11} & 0 \\ 0 & 0 \end{bmatrix} = \begin{bmatrix} D_{11} & 0 \\ 0 & 0 \end{bmatrix} \begin{bmatrix} \bar{N}_{11} & \bar{N}_{12} \\ \bar{N}_{21} & \bar{N}_{22} \end{bmatrix} \tag{2.27}$$

Comparison of submatrices yields

$$\bar{M}_{11}D_{11} = D_{11}\bar{N}_{11}, \quad \bar{M}_{21}D_{11} = 0, \quad D_{11}\bar{N}_{12} = 0. \tag{2.28}$$

Hence, $P^{-1}\bar{M}P$ and $Q^{-1}\bar{N}Q$ are partially reduced, and display r-dimensional components $\bar{M}_{11}$ and $\bar{N}_{11}$ that are similar.

The character χ of a representation is class function whose value χ_k on the conjugacy class C_k with element sum $\bar{C}_k$ is the trace of each of the matrices representing the $°C_k$ elements of the class. Characters of irreducible representations Γ_r have values χ_k^r on class C_k that form a square character table and satisfy the orthogonality relations

$$\sum_k \chi_k^r \bar{\chi}_k^s \, (°C_k/°G) = \delta_{rs} \tag{2.29}$$

$$\sum_r \chi_j^r \bar{\chi}_k^r \, (°C_k/°G) = \delta_{jk} \tag{2.30}$$

If the multiplicities of $\bar{\Gamma}_r$ in $\bar{M}$ and $\bar{N}$ are μ_r^M and μ_r^N then the characters χ^M of $\bar{M}$ and χ^N of $\bar{N}$ satisfy

$$\chi_k^M = \sum_k \mu_r^M \chi_k^r, \quad \chi_k^N = \sum_s \mu_s^N \chi_k^r \tag{2.31}$$

$$\mu_r^M = \sum_k \chi_k^M \bar{\chi}_k^r \, (°C_k/°G) \tag{2.32}$$

$$\sum_k \chi_k^M \chi_k^N \; ({}^oC_k/{}^oG) = \sum_{r,s} \mu_r^M \mu_s^N \sum_k \chi_k^r \chi_k^s ({}^oC_k/{}^oG)$$

$$= \sum_{r,s} \mu_r^M \mu_s^N \delta_{rs} = \sum_r \mu_r^M \mu_r^N = d \qquad (2.33)$$

3. Counting formulas for configuration and modes.

The number d of distinguishable configurations whereby N ligands are attached to N skeletal sites of a molecule is the number of distinct double cosets $\bar{A} \times \bar{B}$, where $\bar{A}$ is the sum of the permutations a of the symmetry group A of the skeleton, and B is the sum of the permutations b of the symmetry group of the ligands (6). It was expressed in terms of induced characters in (2.33). From (2.10) we have for $g \in C_k$

$$\mathrm{tr}\ \mathbf{M}_g = \sum_{i=1}^{m} E(\bar{A}\ r_i g r_i^{-1}) \tag{3.1}$$

$$^{\circ}C_k \chi_k^M = \sum_{i=1}^{m} E(\bar{A}\ r_i \bar{C}_k r_i^{-1}) = m\ E(C_k \bar{A}) \tag{3.2}$$

$$\chi_k^M = (^{\circ}G/^{\circ}C_k) E(\bar{C}_k \bar{A}/^{\circ}\bar{A}) \tag{3.3}$$

Then from (2.33) we have the counting formula

$$d = \sum_k (^{\circ}G/^{\circ}C_k) E(\bar{C}_k \bar{A}/^{\circ}A) E(\bar{C}_k B/^{\circ}\bar{B}) \tag{3.4}$$

The number $^{\circ}G/^{\circ}C_k$ is the order of the centralizer of an element in C_k, which for a permutation of the symmetric group is the product of the lengths of its disjoint cycles times the product of the factorials of the multiplicities of these cycle lengths. In this case (when $\alpha_i = \beta_j = 1$) the number $E(\bar{C}_k \bar{A}/^{\circ}A)$ is the fraction of elements of A in class C_k.

A second counting problem investigated by Hässelbarth and Ruch (5) concerns a classification of equivalent rearrangements x and y for molecules having N different ligands. Disregarding the case where the skeleton is chiral and has no reflectional symmetry σ or where the skeleton is a chiral, but its reflections involve the same site premutations as rotations, we let $\bar{A}$ now denote the sum of the rotational permutations a_i in the normal subgroup A of the complete skeletal symmetry group S, and set

$$\bar{S} = \bar{A} + \bar{A}\sigma = \bar{A}(1 + \sigma) = (1 + \sigma)\ \bar{A}.$$

Rearrangements x and $s\, x\, s^{-1}$ for $s \in S$ are called symmetry equi-

valent, and rearrangements x and ax are called rotationally equivalent. The aggregate $a s x s^{-1}$ for $a \in A$, $s \in S$ and a fixed x in G is an equivalence class $M(x)$ called a mode, whose element sum is the sum of two double cosets

$$\bar{M}(x) = \bar{A} \times \bar{A} + \bar{A}\sigma x \sigma^{-1}\bar{A}.$$

Since $\sigma^2 \in A$, we can write $\sigma\bar{A}$ for $\sigma^{-1}\bar{A}$. The double coset $S \times S$ in G is the union of four double cosets

$$A \times A,\ A\sigma \times \sigma A,\ A\sigma \times A,\ A \times \sigma A$$

which may be either all equal, all distinct, or equal in pairs. Unions of the first two and of the last two form a mode $M(x)$ or $M(\sigma x)$. We present a different, and perhaps simpler method of counting modes than the one published by Hässelbarth and Ruch (5).

<u>Theorem 3.1</u> The number ^{o}M of modes for an N-site skeleton with symmetry group $S = A \dot{+} A\sigma$ containing the rotational subgroup A and a reflection $\sigma \notin A$, is the sum of the numbers of double cosets $\bar{S} \times \bar{S}$ and $\bar{S}^* \times \bar{S}^*$ in G, where $\bar{S}^* = \bar{A}(1-\sigma)$. It is given by the formulas

$$^{o}M = \sum_k (^{o}G/^{o}C_k)\ [E^2(\bar{C}_k\bar{S}/^{o}S) + E^2(\bar{C}_k\bar{S}^*/^{o}S)] \tag{3.5}$$

$$^{o}M = \sum_k (^{o}G/^{o}C_k{}^{o}A^{o}S)[E^2(\bar{C}_k\bar{A}) + E^2(\bar{C}_k\bar{A}\sigma)] \tag{3.6}$$

<u>Proof:</u> Since $\sigma(x\sigma)\sigma^{-1} = \sigma x$, the double cosets $\bar{A}\sigma \times \bar{A}$ and $\bar{A} \times \sigma\bar{A}$ belong to the same mode. If $\bar{A} \times \bar{A}$ is not one of these, then the double coset $\bar{S} \times \bar{S}$ contains two modes, otherwise just one. The weighted double coset

$$\bar{S}^* \times \bar{S}^* = \bar{A}(1 - \sigma) \times (1 - \sigma)\bar{A} \tag{3.7}$$

vanishes if $\bar{A}(1 - \sigma) \times \bar{A} = 0$ or if $\bar{A}\times(1 - \sigma)\bar{A} = 0$, since $\sigma\bar{A} = \bar{A}\sigma$. Otherwise, if $A \times A = A\sigma \times \sigma A$, then

$$\bar{S}^* \times \bar{S}^* = 2\bar{A} \times \bar{A} - 2\bar{A}\sigma \times \bar{A} \tag{3.8}$$

Hence the number of weighted double cosets $\bar{S}^{*} \times \bar{S}^{*}$ counts the number of double cosets $\bar{S} \times \bar{S}$ for which $\bar{A} \times \bar{A}$ is distinct from $\bar{A}\sigma \times \bar{A}$ or $\bar{A} \times \sigma\bar{A}$. Adding the two double coset counts, we obtain (3.5) from (3.4). Writing $\bar{S} = \bar{A}(1 + \sigma)$, $\bar{S}^{*} = \bar{A}(1 - \sigma)$ we have

$$E^2(\bar{C}_k\bar{A}(1 + \sigma) + E^2(\bar{C}_k\bar{A}\ (1 - \sigma) \quad = \quad 2E^2(\bar{C}_k\bar{A}) + 2E^2(\bar{C}_k\bar{A}) + 2E^2(\bar{C}_k\bar{A}\sigma) \tag{3.9}$$

which converts (3.5) to (3.6) and proves the theorem.

4. Vibrational modes and infrared spectra.

If for each of N ligands the three rectangular coordinates of displacement from the equilibrium site are multiplied by the square root of the corresponding mass, the double internal kinetic energy $2E_K$ is expressible in terms of the derivatives of these 3N mass adjusted displacement coordinate x_i by a unit quadratic form (3)

$$2E_K = \sum_{i=1}^{3N} \dot{x}_i^2 = \dot{X}^T\dot{X}, \quad \text{where} \quad \dot{x}_i = dx_i/dt. \tag{4.1}$$

Since the equilibrium configuration $x_i = 0$ represents a minimum of potential energy where $\partial E_i/\partial x_i = 0$ and we can arbitratily choose that minimum value to be 0, the constant and linear terms in the power series expansion of the double potential energy $2E_V$ vanish. Assuming small oscillations about equilibrium we approximate the power series expansion of $2E_V$ by its $3N(3N + 1)/2$ quadratic terms and write

$$2E_V = \sum_{i,j=1}^{3N} v_{ij}x_ix_j + \text{higher order terms} = X^TVX + \ldots \tag{4.2}$$

To simplify the equations of motions, it is desirable to introduce so-called "normal" coordinates q_i by a transformation $X = TQ$, choosing T so that

$$T^TT = I, \quad T^TV\,T = \Lambda = \text{diagonal matrix}. \tag{4.3}$$

Then we have

$$2E_X = \dot{Q}^T\dot{Q}\,, \quad 2E_V = Q^T\Lambda Q = \sum_{i=1}^{3N} \lambda_i q_i^2 \tag{4.4}$$

and the equations of motion become

$$\ddot{q}_i + \lambda_i g_i = 0. \tag{4.5}$$

But how can we choose T to diagonalize an unknown potential energy matrix V?

It seems physically reasonable that the kinetic and potential energy of the system should not be altered if two indistinguishable ligands are interchanged. Such an interchange is a member g of the symmetry group G of the molecule, represented by a matrix S_g operating on the mass adjusted displacement vector X. Furthermore, since only like ligands are interchanged among themselves, S_g can be written as a direct sum of matrices S_g^L operating on the coordinates at sites with ligands of just one kind. Each such component S_g^L is a direct product $R_g x P_g^L$ of a 3 x 3 matrix R_g describing a rigid motion of the skeleton, and a permutation matrix P_g^L of dimension equal to the number of interchangeable ligands of one kind. Since E_K and E_V are unchanged by every S_g we have

$$\dot{X}^T S_g^T S_g \dot{X} = \dot{X}^T \dot{X}, \text{ or } S_g^T S_g = I \qquad (4.6)$$

$$X^T S_g^T VS_g X = X^T VX, \text{ or } S^T VS_g = V, \ VS_g = S_g V. \qquad (4.7)$$

Hence, the unknown matrix V commutes with the symmetry group and must be a linear combination of double coset matrices by Theorem 2.1.

To diagonalize the unknown matrix V, we shall first use double cosets to construct a matrix T such that

$$T^{-1}\bar{S}^L T = \bar{\Gamma} = \sum \bar{\Gamma}_k \times I_{\mu_k^L} \qquad (4.8)$$

is a direct sum of irreducible group matrices. It follows from (4.7) that

$$(T^{-1}VT)\bar{\Gamma} = \bar{\Gamma}(T^{-1}VT) \qquad (4.9)$$

and so, by Schur's lemma cited in Theorem 2.2, $T^{-1}VT$ is a direct sum of blocks of the form

$$T^{-1}VT = \Lambda = \cdot\sum_k I_{f_k} \dot{\times} \Lambda_k \qquad (4.10)$$

where Λ_k is a square matrix whose dimension μ_k^L is equal to the multiplicity of Γ_k in S^L. If all μ_k^L are 0 or 1, T diagonalizes V. Otherwise, additional steps are needed to diagonalize the smaller diagonal blocks Λ_k. Note however, that each eigenvalue of Λ_k appears in **Λ** with multiplicity f_k equal to the degree of Γ_k. When $f_k > I$, the corresponding eigenvalue is sometimes called "degenerate."

We partition the columns of the transforming matrix T into submatrices T_r of f_r columns, corresponding to diagonal blocks Γ_r that are repeated μ_r^L times. Then equation (4.8) implies

$$S^L T_r = T_r \Gamma_r \tag{4.11}$$

Since S^L contains Γ_r with multiplicity μ_r^L, T_r is a linear combination of μ_r^L double coset matrices that intertwine S^L and Γ_r. We can then pick μ_r^L independent intertwining matrices, one for each occurrence of Γ_r, preferably choosing such combinations for which the columns are mutually orthogonal. This can be done if S^L and all Γ_r are orthogonal representations of G, for we then have on transposing equation (4.8) and replacing each group element by its universe that

$$(T^T)\bar{S}^L(T^{-1})^T = \bar{\Gamma} \tag{4.12}$$

so $(T^{-1})^T$ intertwines $\bar{S}^L$ and $\bar{\Gamma}$ and has the same form as T.

Thus, the eigenvectors of the unknown potential energy matrix V are columns of double coset matrices T_r. Furthermore, group theory gives the multiplicities μ_r^L of the eigenvalues in terms of the character χ^L of S^L by a formula like (2.32):

$$\mu_r^L = \sum_k \chi_k^L \chi_k^{(r)} \cdot ({}^oC_k/{}^oG) \tag{4.13}$$

The eigenvalues λ_r are observed from the frequencies $\lambda^{1/2}/2\pi$ of the lines in the molecular spectrum. However, quantum theory predicts overtones of these frequencies not predicted by the classical model described above. Furthermore, selection rules limit the eigenvalues to be observed in the different spectra. Only those frequencies corresponding to irreducible components Γ_r of the 3-dimensional representation called R above, as it operates as a rigid motion on the skeleton, or on the coordinates of the centroid, are active in the infra-red absorption spectrum. Since the permutation representation P^L contains the identity representation Γ_1, it follows that $R \times P^L$ contains R, so there is surely some activity there. The normal coordinates q_i representing the rotations about principal axes of inertia transform according to the 3-dimensional alternating component $R^{[1^2]}$ of $R \times R$, whereas the polarizability tensor transforms according to the 6-dimensional symmetric component $R^{[2]}$. A second selection rule stipulates that only frequencies associated with irreducible components of $R^{[2]}$ are active in the Raman spectrum. If $S^L = R \times P^L$ contains a linear invariant, then P^L contains at least one component of R, and S^L contains a component of $R^{[2]}$. Usually, P^L contains R, so all of $R^{[2]}$ and $R^{[1\ 2]}$ are found in S^L, and active frequencies are obtained in the Raman spectrum. Common frequencies found in the infra-red and Raman spectrum indicate that R and $R^{[2]}$ have common components and may serve to restrict the choice among skeletal models.

5. Application to the methane molecule.

To illustrate the calculation of eigenvectors for the potential energy matrix using double cosets, we consider a model of the methane molecule having four hydrogen sites at the vertices of a regular tetrahedron, and a carbon site at the center. The symmetry group is a direct sum of an irreducible group $S^C = R$ involving rectangular mass-adjusted displacement coordinates x_1, x_2, x_3 for the carbon atom, and a group $S^H = R \times P$ acting on the rectangular mass-adjusted displacement coordinates (displacement $\times\ m^{\frac{1}{2}}$), $x_4, \ldots, x_{15}$ for the four hydrogen atoms. Specifically, if we take the four sites of the regular tetrahedral skeleton to be alternate vertices of a cube, at $(1,1,1)(1 - 1,-1)$, $(-1,1,-1)$, $(-1,-1,1)$ we can take coordinate axes for $x_{3k+1}, x_{3k+3}, x_{3k+3}$ to be parallel respectively to the x,y,z axes through the face centers of the cube.

If the permutation group P is represented by 4×4 permutation matrices that permute four variables y_1, y_2, y_3, y_4, their sum is an invariant. The new vector $z = WY$ of variables, where

$$W = \frac{1}{2}\begin{bmatrix} 1 & 1 & 1 & 1 \\ 1 & 1 & -1 & -1 \\ 1 & -1 & 1 & -1 \\ 1 & -1 & -1 & 1 \end{bmatrix} \tag{5.1}$$

is such that z_1 is an invariant that transforms by the identity representation Γ_1, whereas z_2, z_3, z_4 transform by the irreducible monimial representation R, which we call Γ_4. The tensor product $S^H = R \times P$ is a monomial representation of G acting on the 12 products $z_i y_i$ which correspond to the coordinate variables x_{i+3j-1}. It becomes a permutation representation S^L if we change the signs of the six variables $x_5 = z_3 y_1$, $x_6 = z_4 y_1$, $x_7 = z_2 y_2$, $x_9 = z_4 y_2$, $x_{10} = z_2 y_3$, $x_{11} = z_3 y_3$, making all the 12 hydrogen displacement coordinates increase in outward directions from the central carbon, so that they are permuted among themselves without sign changes under the symmetry operations of G.

The character χ_k^P of P on each class C_k is the number of sites fixed by any permutation of that class and the character of R is

$\chi_k^4 = \chi_k^P - \chi_k^1$ where $\chi_k^1 = 1$. The alternating character χ_m^2 is +1 for even permutations and -1 for odd, and we set $\chi_k^5 = \chi_k^4\chi_k^2$. Finally, $\chi_k^3 = (\chi_k^4)^2 - \chi_k^1 - \chi_k^4 - \chi_k^5$, so we have the character table below.

Class C_k	I	(12) (34)	(123)	(1234)	(12)
${}^oC_k$	1	3	8	6	6
χ_k^1	1	1	1	1	1
χ_k^2	1	1	1	-1	-1
χ_k^3	2	2	-1	0	0
χ_k^4	3	-1	0	-1	1
χ_k^5	3	-1	0	1	-1
χ_k^L	12	0	0	0	2

(5.2)

From the multiplicity formula (2.32) we readily check that

$$\mu_1^L = 1, \ \mu_2^L = 0, \ \mu_3^L = 1, \ \mu_4^L = 2, \ \mu_5^L = 1. \tag{5.3}$$

For the irreducible representations $\Gamma_3, \Gamma_4, \Gamma_5$ of G, we can choose monomial representations induced by linear representations of subgroups of indices 2,3,3. We can also represent S^L as a permutation on the cosets of $A = I + (23)$. Setting

$$r_1 = I, \ r_2 = (14)(23), \ r_3 = (24)(13), \ r_4 = (34)(12) \tag{5.4}$$

$$s_1 = I, \ s_2 = (123), \quad s_3 = (132), \ t = (23)$$

the group matrices and coset representatives in G are

$$\bar{B}_3 = (r_1 + r_2 + r_3 + r_4)(s_1 + \omega s_2 + \omega^2 s_3); \quad 1,T \tag{5.5}$$

$$\bar{B}_4 = (I + t)(r_1 + r_2 - r_3 - r_4); \quad s_j \tag{5.6}$$

$$\bar{B}_5 = (I - t)(r_1 + r_2 - r_3 - r_4) \;; \quad s_j \tag{5.7}$$

$$\bar{A} = (I + t) \;; \quad r_i s_k \tag{5.8}$$

Here $\omega^3 = 1$, and $\bar{B}_5$ is obtained from $\bar{B}_4$ by changing the signs on odd permutations $t\, r_i$.

To calculate a 12×12 matrix T that transforms $\bar{S}^L$ into a direct sum of $\bar{\Gamma}_1$, $\bar{\Gamma}_3$, $\bar{\Gamma}_4$, $\bar{\Gamma}_4$, $\bar{\Gamma}_5$, we partition the columns in sets of 1,2,3,3,3 corresponding to the degrees of the Γ_r. Each submatrix T_j of T is a double coset matrix that intertwines the group matrix of G induced by $\bar{A}$ with the group matrix induced by a $\bar{B}_r$.

If we reorder the rows of T according to coset representatives placing $r_i s_k$ in row $i + 4(k - 1)$, so the four x-displacements at hydrogen sites precede the four y-displacements which precede the z-displacements, the 12 entries in a column of T can be written as multiples of 3 columns of the matrix W in [5,1]. Three vectors W_1 form the column T_1 in (5.11) that intertwines $\bar{S}^L$ with $\bar{\Gamma}_1$ and has equal entries.

There is only one double coset $\bar{A}\bar{B}_3$, since every element of G is a product of I or t in A by an even permutation in $\bar{B}_3$. Columns 1 and 2 of the double coset matrix T_2 are the complex conjugates of the coefficients in $\bar{A}\bar{B}_3$ of $r_i s_k$ and $r_i s_k t^{-1} = r_i t\, s_k^{-1}$ respectively.

Since $\bar{\Gamma}_4$ is contained twice in $\bar{S}^L$, there are two double cosets of $\bar{A}$ and $\bar{B}_4$, namely

$$\bar{A}\bar{B}_4 = 2\bar{B}_4, \text{ and } \bar{A}s_2\bar{B}_4 = (s_2+s_3)\bar{B}_4 = \bar{A}s_3\bar{B}_4. \tag{5.9}$$

If T_3 is chosen to be the double coset matrix for $\bar{A}\bar{B}_4$, the corresponding quasi-normal coordinates q_i will be those for the centroid of the four hydrogen atoms. The normal coordinates q_i related to the double coset matrix T_4 for $\bar{A}s_2\bar{B}_4$ will represent vibrations within the hydrogen system. Coefficients in row $i\,k$, column j, of T_3 and T_4 are respectively the complex conjugates of the coefficients of $r_i s_k s_j^{-1}$ in $\bar{A}\bar{B}_4$ and $\bar{A}s_2\bar{B}_4$. Since $(I + t)(I - t) = 0$, the double coset $\bar{A}\bar{B}_5$ vanishes, and the double coset whose matrix is T_5 is

$$\bar{A}s_2\bar{B}_5 = (s_2 - s_3)\bar{B}_5 = -\bar{A}s_3\bar{B}_5 \tag{5.10}$$

We can now express T in the form

$$T = \left[\begin{array}{c|cc|ccc|ccc|ccc} W_1 & W_1 & W_1 & W_2 & 0 & 0 & 0 & W_3 & W_4 & 0 & -W_3 & W_4 \\ W_1 & \bar{\omega}W_1 & \omega W_1 & 0 & W_2 & 0 & W_4 & 0 & W_3 & W_4 & 0 & -W_3 \\ W_1 & \omega W_1 & \bar{\omega}W_1 & 0 & 0 & W_2 & W_3 & W_4 & 0 & -W_3 & W_4 & 0 \end{array}\right] \tag{5.11}$$

$$T_1 \qquad T_2 \qquad T_3 \qquad T_4 \qquad T_5$$

The columns of T are mutually orthogonal, since $W_i^T W_j = \delta_{ij}$. T becomes an orthogonal matrix if we divide each column in (5.11) by its length, namely $\sqrt{3}$ for T_1 and T_2 and $\sqrt{2}$ for T_4 and T_5. A permutation of the rows of T will restore the order of the origin variables x_4 to x_{15}, yielding T_H and we replace T by $T_c \dotplus T_H = T_M$ to include the 3 x 3 identity matrix T_c that intertwines the irreducible carbon atom representation $S^c = R$ with an additional copy of $\Gamma_4 = R$.

To completely diagonalize $T_M^{-1} V T_M$ we replace the submatrices T_c and T_3, associated with Γ_4 and the mass adjusted displacement coordinates for the carbon and hydrogen centroids, by two mutually orthogonal linear combinations, the first (T_0) yielding mass adjusted displacement coordinates for the centroid of the molecule, and the other T_3^* involving the relative motion of the carbon and hydrogen centroids:

$$T_0 = T_0 \cos\varphi + T_3 \sin\varphi, \quad T_3^* = -T_0 \sin\varphi + T_2 \cos\varphi, \quad \tan^2\varphi = 4m/\mathbf{M} \tag{5.12}$$

where $m/\mathbf{M}$ is the hydrogen carbon mass ratio.

Thus, using double coset matrices, we have found in the columns of $T_{\mathbf{M}}$ a complete set of eigenvectors for the completely unknown potential energy matrix V. To find the eigenvalues λ_i and compute $V = T_{\mathbf{M}} \Lambda T_{\mathbf{M}}^T$ we examine the molecular spectrum.

The three equal eigenvalues λ_0 associated with the representation Γ_4 for the coordinates of the molecular centroid, and the three equal eigenvalues λ_5 associated with the representation Γ_5 for the components of angular velocity must vanish in the absence of external forces and moments, since the corresponding vibration equation $d^2q_i/dt^2 + \lambda_i q_i = 0$ in (4.5) must yield unaccelerated motion. There remain four distinct eigenvalues $\lambda_1, \lambda_2, \lambda_3, \lambda_4$ of multiplicities 1,2,3,3 associated with the nine internal degrees of freedom. The two frequencies with eigenvalues λ_3 and λ_4 associated with Γ_4 are infra-red active, while those with λ_1 and λ_2 are not. Since

$$R^{[2]} \cong \Gamma^{(1)} \dot{+} \Gamma^{(3)} \dot{+} \Gamma^{(4)}$$

all four frequencies λ_i are active in the Raman spectrum. These four fundamental frequencies predicted by the classical theory must be augmented by overtones predicted by the quantum theory to obtain a complete spectral analysis.

We have considered the consequences of assuming a tetrahedral skeleton for methane. If a square skeleton were assumed, the symmetry group would have a subgroup of eight rotations, with a coset containing a reflection in the skeletal plane. This group has eight linear and two 2-dimensional irreducible representations. We find one simple and two double infra-red frequencies, none of which are Raman active, three other simple Raman frequencies, and one mode not active in either. Thus, the spectral patterns can distinguish the tetrahedral from the square model and verify the former.

References

1. Frame, J.S., The double cosets of a finite group, Bull. A.M.S. v. 47, (1941), pp. 458-467.

2. Frame, J.S., Group decomposition by double coset matrices, Bull. A.M.S. v. 54 (1948) 740-755.

3. Frame, J.S., Symmetry groups and molecular structure, Pi Mu Epsilon J. (1959) pp. 463-471.

4. Frame, J.S., The constructive reduction of finite group representations, A.M.S. Proc. of Symposia in Pure Math., 1960 Institute of Finite Groups, Vol. 6 (1962) 89-99.

5. Hässelbarth, W., Ruch, E., Classifications of Rearrangement Mechanisms by means of Double Cosets and Counting Formulas for the Numbers of Classes. Theoret. Chim. Acta (Berl.) 29, 259-268 (1973).

6. Ruch, E., Hässelbarth W., and Richter, B., Doppelnebenklassen als Klassenbegriff und Nomenklaturprinzip für Isomere und ihre Abzählung, Theoret. Chim. Acta (Berl.) 19, 288-300 (1970).

The Chirality Algebra

Some comments concerning mathematical aspects of the Ruch/Schönhofer chirality theory

by Andreas W.M. Dress, Bielefeld

Dedicated to Ernst Ruch on the occasion of his 60th birthday

Introduction: The purpose of the Ruch/Schönhofer chirality theory.

An important branch of stereochemistry is the study of pseudoscalar observations on molecules, i.e. of observations which - like optical activity - have the same absolute value but different sign for substances consisting of molecules which are mirror images of each other.

Following Ernst Ruch and his collaborators we want to do this by comparing the results of such observations on large classes of molecules which have basically the same spatial and molecular structure but varying ligands at certain specified sites. [1)] It is the purpose of the Ruch/Schönhofer theory of chirality functions to provide the mathematical machinery necessary to analyse such data in terms of (a) the spatial and modecular structure of our molecular class and (b) the chemical structure of the varying ligands.

The symmetry of the spatial and molecular structure is used to restrict considerably the structure of those functions which - depending on the chemical structure of the varying ligands as their variables - are assumed to describe the pseudoscalar observation we started with. There is some good reason for the hope that this way it will be possible to determine at the same time the specific chemical property of the varying ligands (in terms of a "ligand parameter") which is responsible for what is being observed as well as the specific (interacting) forces which are characteristic for the spatial and molecular structure of our molecular class and which determine the specific form of the "chirality functions", i.e. the function whose variables are the "ligand parameters"

1) see [11] One such class is for instance the class of methane derivatives, studied in [12] and [15], another one the class of allene derivatives studied in [13] and [14].

and whose values correspond with the experimental pseudoscalar observation.

It is the purpose of this paper to popularize the Ruch/Schönhofer theory among mathematicians by developing it from a set of rather formal definitions. I will not justify each definition individually, not only, to keep this paper short, but also, since I believe that the only appropriate justification of a theory has to rely on the w h o l e b o d y of the theory (its definitions, concepts, results etc. altogether) as a tool to understand the specific phenomena of reality the theory deals with, and not so much on its individual definitions.

1. Notations and Definitions.

Let $\mathbb{E}^3$ denote the three dimensional euclidean space (which - by using a cartesian coordinate system- can be identified with $\mathbb{R}^3$, the set of tripels (x_1,x_2,x_3) of real numbers), let $d(P,Q)$ denote the distance of P and Q for $P,Q \in \mathbb{E}^3$ ($d(P,Q) = \sqrt{(x_1-y_1)^2+(x_2-y_2)^2+(x_3-y_3)^2}$, if P is identified with (x_1,x_2,x_3) and Q with (y_1,y_2,y_3)) and let $O(\mathbb{E}^3)$ denote the group of all distance preserving maps: $\alpha : \mathbb{E}^3 \to \mathbb{E}^3$ of $\mathbb{E}^3$ into itself:

$$O(\mathbb{E}^3) = \{\alpha : \mathbb{E}^3 \to \mathbb{E}^3 \mid d(\alpha(P),\alpha(Q)) = d(P,Q) \text{ for all } P,Q \in \mathbb{E}^3\}$$

$O(\mathbb{E}^3)$ is wellknown to be the semi direct product of the normal subgroup $T \simeq \mathbb{R}^3$ of
$O(\mathbb{R}^3) \simeq O(\mathbb{E}^3)/T$ with respect to its natural action on $T \simeq \mathbb{R}^3$. For every $P \in \mathbb{E}^3$ the stabilizer group $O(\mathbb{E}^3)_P = \{\alpha \in O(\mathbb{E}^3) \mid \alpha(P) = P\}$ is isomorph to $O(\mathbb{R}^3)$.

For any $\alpha \in O(\mathbb{E}^3)$ denote by $\det \alpha$ the "determinant" of α, i.e. the determinant of the image $\alpha + T$ of α in $O(\mathbb{E}^3)/T \simeq O(\mathbb{R}^3)$, and let $O^+(\mathbb{E}^3) = \{\alpha \in O(\mathbb{E}^3) \mid \det \alpha = 1\}$.

One has

$$\det(\alpha\cdot\beta) = \det \alpha \cdot \det \beta,$$
$$\det \alpha = \pm 1$$

and

$$\det \alpha = \begin{cases} +1, & \text{if } \alpha \text{ preserves the "orientation" of } \mathbb{E}^3 \\ -1, & \text{if } \alpha \text{ reverses the "orientation" of } \mathbb{E}^3, \end{cases}$$

thus $O^+(\mathbb{E}^3) = \mathrm{ke}(\det)$ is a normal subgroup of $O(\mathbb{E}^3)$ of index 2.

Now we define a (rigid [2]) skeleton $S = (S,P)$ to be an arbitrary subset $S \subseteq \mathbb{E}^3$ together with a finite subset $P \subseteq S$, the set of "sites" of S, at which ligands may be fixed [3]. The symmetry group of our skeleton (S,P) is defined to be

$$G = G_S = \{\alpha \in O(\mathbb{E}^3) | \alpha(S) = S,\ \alpha(P) = P\}.$$

Since any $\alpha \in G$ maps the finite set P onto itself, it fixes the center of gravity C of P. Thus G cannot contain any translation: $G \cap T = \mathrm{Id}_{\mathbb{E}^3}$, and can be considered as a subgroup of $O(\mathbb{R}^3) \simeq O(\mathbb{E}^3)_C$.

We define

$$N = N_S = \{\alpha \in G_S | \det \alpha = +1\} = G_S \cap O^+(\mathbb{E}^3)$$

to be the proper symmetry group of S.

Since any $\alpha \in G$ maps the finite set P onto itself, it can be considered as a permutation of P. Thus we have a homomorphism

$$G \to \Sigma_P : \alpha \to \bar{\alpha} = \alpha|_P ,$$

where Σ_P denotes the full permutation group of the finite set P ($\Sigma_P \simeq \Sigma_n$, the nth symmetry group, if P contains precisely n elements) and $\bar{\alpha}$ the restriction of α to P. Let $\bar{G} = \{\bar{\alpha} \in \Sigma_P | \alpha \in G\}$ and $\bar{N} = \{\bar{\alpha} \in \Sigma_P | \alpha \in N\}$ denote the image of G and N, respectively, in Σ_P with respect to this homomorphism. The only interesting situation for the Ruch/Schönhofer theory occurs in case $\bar{G} \neq \bar{N}$, thus we assume this in the following. It means that S is achiral (i.e. it coincides - up to proper rotations - with its mirror image or, in other words, there exists an improper map $\alpha \in G \setminus N$) and that any improper map $\alpha \in G \setminus N$ induces a non-trivial permutation $\bar{\alpha}$ on P. It implies that $(G:N) = (\bar{G}:\bar{N}) = 2$.

Now we assume a finite set $L = \{\ell_1,\ldots,\ell_t\}$ of "ligands" being given. A molecule m - with respect to S and L - is then defined to be a

---

2) For "non-rigid" molecules see § 5

3) In case of methane derivatives S may be choosen either as a regular tetrahedron or as the union of the edges of this tetrahedron or as the union of the straight lines, connecting the center of the tetrahedron with its corners, or ... and P as the set of the four corners.

(set-theoretic) map $m: P \to L$. If $m(P) = \ell_i$, we interpret this as "the ligand ℓ_i is attached to S at the site P". The set $M = M(S,L) = \{m: P \to L\} = L^P$ of all (set-theoretic) maps of P in L represents the set (or class) of all molecules m with skeleton $S = (S,P)$ and ligands from L.

We observe that Σ_P acts naturally from the right on M: for $m \in M$ and $\sigma \in \Sigma_P$ we define $m\sigma: P \to L$ by:

$$(m\sigma)(P) = m(\sigma(P)),\ P \in P.$$

We define two "molecules" $m, m' \in M$ to be "isomers", if there exists a permutation $\sigma \in \Sigma_P$ with $m\sigma = m'$. m and m' are defined to be "rotation equivalent", if there exists some $\alpha \in N$ with $m\bar{\alpha} = m'$, and they are defined to be "enantiomers", if there exists $\alpha \in G-N$ with $m\bar{\alpha} = m'$.

2. The space of chirality observations and the Ruch chirality algebra.

After having defined the set - or class - of molecules with skeleton S and ligands in L we now define a pseudoscalar or "chirality" observation on this set $M = M(S,L)$ to be a map

$$F : M \to \mathbb{R},$$

which associates with any molecule $m \in M$ a real number - the result of our observation - such that

$$F(m\bar{\alpha}) = \det \alpha \cdot F(m)$$

for all $m \in M$ and $\alpha \in G$, i.e. $F(m)$ and $F(m')$ coincide for "rotation-equivalent" molecules m and m' and have opposite sign, but the same absolute value for enantiomers.

Obviously the sum $F_1 + F_2$ of two chirality observations F_1 and F_2 is again a chirality observation and so is $c \cdot F$ for any $c \in \mathbb{R}$ and chirality observation F. Thus the set of all chirality observations - on molecules with skeleton $S = (S,P)$ and ligands in L - forms a vector space $Ob(S,L)$.

To characterize this vector space as a subspace of the vector space $\mathbb{R}^M$

of all maps from M to $\mathbb{R}$, (i.e. of all possible observations on M, whether they are chirality observations or not) let us observe, that the (right) action of Σ_P on M induces a (left) action of Σ_P on $\mathbb{R}^M$, defined by

$$(\sigma F)(m) = F(m\sigma) \quad (\sigma \in \Sigma_P, F \in \mathbb{R}^M, m \in M).$$

This action is obviously linear $(\sigma(F_1 + F_2) = \sigma F_1 + \sigma F_2, \sigma(c\cdot F) = c\cdot\sigma(F))$ and thus it can be extended naturally to an action of the full group ring $\mathbb{R}\Sigma_P = \{\sum_{\sigma\in\Sigma_P} a_\sigma\sigma | a_\sigma \in \mathbb{R}\}$ on $\mathbb{R}^M$:

$$((\sum_{\sigma\in\Sigma_P} a_\sigma\sigma)\cdot F)(m) = \sum_{\sigma\in\Sigma_P} a_\sigma F(m\sigma),$$

(with $\Sigma a_\sigma\sigma \in \mathbb{R}\Sigma_n$, $F \in \mathbb{R}^M$, $m \in M$).

Now consider the specific element $\chi_S = \frac{1}{|G|} \sum_{\alpha\in G} \det \alpha\cdot\bar{\alpha} \in \mathbb{R}\Sigma_P$, which is called the "chirality operator", associated with S. We have

$$\alpha\cdot\chi_S = \det \alpha\cdot\chi_S$$

for all $\alpha \in G$ and thus $\chi_S^2 = \chi_S$, i.e. χ_S is an idempotent in the group ring $\mathbb{R}\Sigma_P$. Moreover, one can easily verify:

<u>Proposition 1:</u> $Ob(S,L) = \{F:M \to \mathbb{R} \in \mathbb{R}^M | \chi_S F = F\} = \chi_S\,\mathbb{R}^M$.

The purpose of the following notes is to split $Ob(S,L)$ into "irreducible components" with respect to our symmetry operations, thus splitting any chirality observation into a sum or "superposition" of "primitive" or "indecomposable" chirality observations which are supposed to correspond chemically to independent chirality phenomena.

The first problem, of course, is to specify those symmetry operations which act <u>naturally</u> on $Ob(S,L)$, so that it makes sense, chemically and mathematically, to split $Ob(S,L)$ into irreducible components with respect to this action. One can neither take all operators in the symmetric group ring $\mathbb{R}\Sigma_P$ nor all in the group ring $\mathbb{R}G$ or $\mathbb{R}\bar{G}$ because - in general - there will be operators $\psi \in \mathbb{R}\Sigma_P$ and even in $\mathbb{R}\bar{G}$ which do not fix $Ob(S,L) = \chi_S\cdot\mathbb{R}^M$ (as a whole), since $\psi\cdot(\chi_S\cdot F)$ might not be of the form $\chi_S\cdot F'$. But rather than looking around for some appropriate definition to come along, one should look at the problem and then one may realize that an appropriate operator algebra to act on $Ob(S,L)$ is, of course, the set of all elements $\psi \in \mathbb{R}\Sigma_P$ which are of the form $\chi_S \cdot \psi' \cdot \chi_S$ for some $\psi' \in \mathbb{R}\Sigma_P$ (and thus of the form $\chi_S \cdot \psi \cdot \chi_S$, since $\chi_S^2 = \chi_S$). Thus, we define the Ruch chirality algebra, associated with our skeleton S (cf. [11], p. 240/1), to be

$$Ch(S) = \chi_S \cdot \mathbb{R}\Sigma_P \cdot \chi_S = \{\psi \in \mathbb{R}\Sigma_P | \chi_S \cdot \psi = \psi\chi_S = \psi\}.$$

Remark: Of course, one might also consider the larger algebra $Ch'(S)$, consisting of all $\psi \in \mathbb{R}\Sigma_P$, for which there exists some $\psi' \in \mathbb{R}\Sigma_P$ with $\psi \cdot \chi_S = \chi_S \cdot \psi'$, since also those will map $Ob(S,L) = \chi_S \cdot \mathbb{R}^M$ into itself. But since we are interested in this algebra only with respect to its action on $Ob(S,L) = \chi_S \cdot \mathbb{R}^M$ and thus have to consider this algebra only modulo those $\psi \in Ch'(S)$, for which $\psi \cdot \chi_S = 0$, we can use the natural isomorphism $Ch'(S)/\{\psi \in Ch'(S) | \psi \cdot \chi_S = 0\} \cong Ch(S)$, induced by $\psi \mapsto \psi\chi_S = \chi_S\psi\chi_S$ (the last equality follows from $\psi\chi_S = \chi_S\psi'$!), and are back at $Ch(S)$.

Now we use the following wellknown facts, which are easily verified:

Proposition 2: Let A be a semi-simple algebra and let $e \in A$ be an idempotent ($e^2 = e$). Then eAe is a semi-simple algebra with e as its unit.

Let $e_1,\ldots,e_t$ be the primitive central idempotents of A and let $M_1,\ldots M_t$ be the corresponding irreducible representation spaces of A (i.e. A-modules) (thus $e_i \cdot m_j = \delta_{ij} m_j$ for $m_j \in M_j$). Then the primitive central idempotents of eAe are precisely those products $e \cdot e_i = e \cdot e_i \cdot e$ which are non-zero ($i \in \{1,\ldots,t\}$) and the corresponding irreducible representation spaces of eAe are the spaces eM_i (of course, $eM_i \neq 0 \Leftrightarrow e \cdot e_i \neq 0$).

Finally, for any decomposition of e in A into pairwise orthogonal (primitive) idempotents $e = f_1 + \ldots + f_s$ the constituents f_j are contained in eAe, thus any primitive idempotent eAe is primitive as well in A and two primitive idempotents f_1 and f_2 in eAe belong to the same primitive central idempotent ee_i in eAe (i.e. $(ee_i) f_1 \neq 0 \neq (ee_i) f_2$) if and only if they belong to the same primitive central idempotent e_i in A, (i.e. $e_i f_1 \neq 0 \neq e_i f_2$).

Thus the representation theory of eAe can be read off from the representation theory of A.

Proof: Obviously $e \in eAe$ is the unit element in eAe. If M is an A-representation space, then eM is an eAe-representation space. If $M_1 \subseteq eM$ is an eAe-invariant subspace of eM, then $A\cdot M_1 \cap eM = M_1$, because for any element $\sum_{i=1}^{k} a_i m_i \in A\cdot M_1$ $(a_i \in A,\ m_i \in M_1 \subseteq eM)$ with $\sum_{i=1}^{k} a_i m_i \in eM$ one gets:

$$\sum_{i=1}^{k} a_i m_i = e\cdot(\sum_{i=1}^{k} a_i(em_i)) = \sum_{i=1}^{k} (ea_ie)m_i \in (eAe)\cdot M_i = M_i .$$

Thus eM is irreducible if M is, eA is a faithful semi-simple representation space of eAe, so eAe is semi-simple and the irreducible eAe-representation spaces are precisely those of the form $e\cdot M_i$, which are non zero. But $e\cdot M_i \neq 0 \Leftrightarrow (e\cdot e_i)\cdot M_i \neq 0$ and $ee_i\big|_{eM_j} = \delta_{ij}\ \mathrm{Id}\big|_{eM_j}$ for all i,j, thus ee_i is the primitive central idempotent in eAe corresponding to eM_i, in particular $ee_i \neq 0 \Leftrightarrow eM_i \neq 0$.

Finally, if $e = f_1 + \ldots + f_s$ with $f_if_j = \delta_{ij}f_j$ and $f_i \in A$, then $ef_j = (\sum_i f_i)f_j = \sum_i \delta_{ij}f_j = f_j$ and similarly $f_j\cdot e = f_j$, thus $f_j = ef_je \in eAe$.

So, from Prop. 2, we get a 1-1-correspondance between the irreducible representation spaces of $Ch(S) = \chi_S\, \mathbb{R}\Sigma_S\, \chi_S$ and some irreducible representation spaces of $\mathbb{R}\Sigma_P$. But those are wellkown to correspond in a 1-1-fashion to the partitions $\gamma = (n_1,\ldots,n_r)$ of $n = |P|$, i.e. to the indexed families of natural numbers $n_1,\ldots,n_r \in \mathbb{N}$ of varying length r with $\sum_{\rho=1}^{r} n_\rho = n$ and $n_1 \geq n_2 \geq \ldots \geq n_r$ (obviously $r \leq n$). Let N_γ denote the corresponding irreducible representation space of $\mathbb{R}\Sigma_P$ and let e_γ denote the corresponding primitive central idempotent, as constructed by Young (see for instance [11], Anhang A). Then the non zero products $\chi_S\cdot e_\gamma = \chi_{S,\gamma}$ are the primitive central idempotents of $Ch(S)$ and the spaces $\chi_S\cdot N_\gamma = N_{S,\gamma} \subseteq N_\gamma$ represent the various irreducible $Ch(S)$-spaces.

According to Ruch and Schönhofer we define γ to be an active partition with respect to S if and only if $\chi_S\cdot e_\gamma = \chi_{S,\gamma} \neq 0$. Thus the representation theory of Ch(S) can be described in terms of the active partitions γ.

3. The "phase portrait" of a chirality observation.

Let $F \in Ob(S,L)$ be a chirality observation. We define F to be of type γ (for some active partition γ) if $\chi_{S,\gamma}\cdot F = F$. Any F splits uniquely into a sum of chirality observations of the various types γ:

$$F = \chi_S \cdot F = (\sum_\gamma \chi_{S,\gamma})\cdot F = \sum_\gamma(\chi_{S,\gamma}\cdot F)$$

and $F = \sum_\gamma F_\gamma$ with $\chi_{S,\gamma}\cdot F_\gamma = F_\gamma$ implies

$$\chi_{S,\gamma_0}\cdot F = \sum_\gamma \chi_{S,\gamma_0} F_\gamma = \sum_\gamma \chi_{S,\gamma_0}\cdot \chi_{S,\gamma}\cdot F_\gamma = \sum_\gamma \delta_{\gamma_0,\gamma} F_\gamma = F_{\gamma_0} .$$

Moreover, we may consider the Ch(S)-submodule of $Ob(S,L)$ generated by F, i.e. $Ch(S)F = \{\psi\cdot F | \psi \in Ch(S)\} \subseteq Ob(S,L)$ and split it into irreducible Ch(S)-spaces. The number $\mu_\gamma(F)$ of components of such a splitting which are isomorphic to $N_{S,\gamma}$ is independent of the splitting and an invariant of F. It is necessarily smaller or equal to $n_{S,\gamma} = \mathrm{Dim}_{\mathbb{R}}\, N_{S,\gamma}$ (since there are only that many components isomorphic to $N_{S,\gamma}$ in any splitting of Ch(S) and $Ch(S)\cdot F$ is a homomorphic image and thus isomorphic to a direct summand of Ch(S). Moreover, $n_{S,\gamma}\cdot\mu_\gamma(F) = \mathrm{Dim}_{\mathbb{R}}(e_\gamma\cdot Ch(S)F)$. We define the family of numbers $\{\mu_\gamma(F)|\gamma$ active for $S\}$ the "phase portrait" of F. It gives the number of irreducible components (chirality phenomena) of type γ contained in F (or rather in $Ch(S)\cdot F$). It is an important question whether or not there are a priori arguments (quantum chemically or otherweise) which imply that $\mu_\gamma(F) = 0$ for certain active partitions γ and certain specified classes of chirality phenomena F.

There is an amazing result in this direction by Ruch and Schönhofer ([11], see also [2]), which is based on an ingenious interpretation of the fact, that the irreducible representation spaces of Ch(S) and of $\mathbb{R}\Sigma_P$ are indexed by partitions of $n = |P|$: for any molecule $m:P \to L$ we can define its associated "ligand partition" γ_m by considering for any ligand $\ell \in L$ the number m_ℓ of sites $P \in P$, at which ℓ is being fixed: $m_\ell = |\{P \in P | m(P) = \ell\}| = |m^{-1}(\ell)|$ and by ordering these numbers according to their size, e.g. $m_{\ell_1} \geq m_{\ell_2} \geq \ldots \geq m_{\ell_r} \neq m_{\ell_{r-1}} = \ldots = m_{\ell_t} = 0$, which defines a partition $\gamma_m = (m_{\ell_1},\ldots,m_{\ell_r})$ of $n = |P|$ because of

$$\sum_{\rho=1}^{r} m_{\ell_\rho} = \sum_{\rho=1}^{t} m_{\ell_\rho} = \sum_{\rho=1}^{t} |m^{-1}(\ell_\rho)| = |\bigcup_{\rho=1}^{t} m^{-1}(\ell_\rho)| = |m^{-1}(L)| = |P|.$$

We may now restrict ourself to molecules m with a prescribed ligand partition $\gamma_m = \gamma$. Any isomer m' of m defines the same ligand partition, thus the set $M_\gamma = \{m \in M = L^P | \gamma_m = \gamma\}$ of all molecules with ligand partition γ is invariant as a whole under the action of Σ_P.

Therefore the space $\chi_S \cdot \mathbb{R}^{M_\gamma} = \mathrm{Ob}(S,L;\gamma)$ of chirality observations on molecules with prescribed ligand partition γ is a well defined representation space for $\mathrm{Ch}(S)$.

The question is now whether for a chirality observation $F \in \mathrm{Ob}(S,L;\gamma)$ the set of active partitions γ' with $\chi_{S,\gamma'} \cdot F = F_{\gamma'} \neq 0$ is somehow restricted by γ, i.e. whether the types of chirality observations on M_γ exclude certain, otherwise active types.

The central result of Ruch and Schönhofer is the following

<u>Theorem</u>: For any two partitions $\gamma = (n_1,\ldots,n_r)$ and $\gamma' = (m_1,\ldots,m_s)$ of n define the relation "$\gamma \geq \gamma'$" if and only if for all $t = 1,\ldots,\min(r,s)$ one has

$$\sum_{\rho=1}^{t} n_\rho \geq \sum_{\rho=1}^{t} m_\rho .$$

Then there exists a chirality observation $0 \neq F \in \mathrm{Ob}(S,L;\gamma)$ of type γ', i.e. with $\chi_{S,\gamma'} \cdot F = \ell_{\gamma'} \cdot F = F$, if and only if $\gamma \geq \gamma'$.
A proof of this result can be found in [11] (see also [2]). It is more or less equivalent with a number of other non trivial combinatorial theorems on partition diagrams, matrix theory and the representation theory of symmetric groups. There is also a number of interesting papers of Ruch et al. on the significance of the order relation $\gamma \geq \gamma'$, in particular with respect to the notion of entropy and the so called Boltzmann H-theorem, see [6], [9] and [10].

Let us finally consider the physical significance of $\mathrm{Ch}(S)$ and its action on $\mathrm{Ob}(S,L)$ and $\mathrm{Ob}(S,L;\gamma)$ which so far has been studied only from a rather formal point of view.

Following Ruch and Schönhofer we first define "mixtures" of molecules as formal sums $\sum_{m\in M} x_m \cdot m$ with $x_m \in \mathbb{R}$. We interpret this as a mixture, e.g. in a solution, with $|x_m|$ denoting the concentration of the substance consisting of molecules of type m, if $x_m \geq 0$, and of type m' for some enantiomer m' of m, if $x_m < 0$. Again one can define a right action action of Σ_P and even of $\mathbb{R}\Sigma_P$ on the "space of mixtures" by putting

$$\left(\sum_{m\in M} x_m \cdot m\right)\left(\sum_{\sigma\in\Sigma_P} a_\sigma \sigma\right) = \sum_{m,\sigma} x_m a_\sigma (m\sigma).$$

Also, two "mixtures" $\sum_{m\in M} x_m m$ and $\sum_{m\in M} y_m m$ may describe the same mixture, physically, e.g. $\sum_{m\in M} x_m m$ and $\sum_{m\in M} (\det \alpha)\cdot x_m (m\bar{\alpha})$ is the "same" mixture, physically, for any $\alpha \in G_S$. More precisely one has (similarly to Prop. 1)

Proposition 3: Two mixtures $\sum_{m\in M} x_m m$ and $\sum_{m\in M} y_m m$ define the same mixture, physically, if and only if

$$\sum_{m\in M} x_m m\cdot \chi_S = \sum_{m\in M} y_m m\cdot \chi_S .$$

In particular, the mixture $\sum_{m\in M} x_m m\cdot \chi_S$ defines the same mixture as $\sum x_m \cdot m$ and thus the space

$$\mathrm{Mix}(S,L) = \{\sum_{m\in M} x_m \cdot m \,|\, x_m \in \mathbb{R}\}\cdot \chi_S$$

represents all physical mixtures in a 1-1-fashion.

Ch(S) acts on Mix(S,L) from the right as a prescription of how to produce some other (though isomeric) mixture from a given one.

There is a canonical pai ring of Mix(S,L) and Ob(S,L), given by

$$\langle \sum x_m m, F\rangle = \sum x_m F(m),$$

which identifies Ob(S,L) with the dual of Mix(S,L) and any chirality observation F with a linear functional on Mix(S,L). For any $\psi \in$ Ch(S) one has obviously

$$\langle (\sum x_m \cdot m)\psi, F\rangle = \langle \sum x_m \cdot m, \psi\cdot F\rangle$$

for all $\Sigma\, x_m m \in Mix(S,L)$ and $F \in Ob(S,L)$. Thus one may interpret $\psi \cdot F$ as a chirality observation on our mixture $\Sigma\, x_m m$, which is produced by observing F on the new mixture $(\Sigma\, x_m \cdot m) \cdot \psi$, we get from $\Sigma\, x_m \cdot m$ by using our prescription ψ. The space $Ch(S) \cdot F \subseteq Ob(S,L)$ of chirality observations, generated by F in this way, is irreducible if and only if any non-zero product $\psi \cdot F$ regenerates F, i.e. there is some $\psi' \in Ch(S)$ with $\psi'(\psi \cdot F) = F$. Otherwise we can split F, i.e. we can find elements (even primitive idempotents) $\psi_1, \ldots, \psi_s \in Ch(S)$ such that $\psi_i \cdot F = F_i$ is irreducible (i.e. generates an irreducible $Ch(S)$ representation space $Ch(S) \cdot \psi_i\, F$), the F_i are independent (i.e. a non-zero product $\psi\, F_i$ can not be represented in the form $\sum_{j \neq i} \psi_j F_j ; \psi, \psi_j \in Ch(S)$) and $F = F_1 + \ldots + F_s$.

This way it becomes obvious that the physical significance of the above method of analyzing a given chirality phenomenon depends on the physical significance of the above procedures to produce new mixtures from given ones according to the prescriptions $\psi \in Ch(S)$. Though, at first, this might look rather formal and, may be, as artificial as the original formal definition of the action of $Ch(S)$ on $Ob(S,L)$, it should be kept in mind, that the prescription $\psi \in Ch(S)$, in particular in the most important case of a primitive idempotent $\psi = \psi_i = \psi_i^2$, may help to produce mixtures, in which certain, if not all but one chiral forces are cancelled out, so that we are left with less or only just one isolated chiral force, which is observed by $F_i = \psi_i F$, whereas F as a whole is just the result of the superposition of those "chiral forces" $F = F_1 + \ldots + F_s$.

5. Chirality Functions.

In the following I want to explain how chirality observations can be modeled by chirality functions. So let $\{x_P | P \in P|$ be a set of $|P| = n$ indeterminates or variables and consider the ring $\mathbb{R}[x_P | P \in P]$ of all real polynomials in these variables.

Remark: It would be more conventional to index the elements in P by natural numbers: $P = \{P_1, \ldots, P_n\}$, and then to consider polynomials in $x_1 = x_{P_1}, \ldots, x_n = x_{P_n}$. But since it is essential in this (as in any other) subject to have full control about transformation laws etc. (what acts on what from which side ... ?) and not confuse, say, the indices of the sites

$P_1,\ldots,P_n$ and those of the ligands $\ell_1,\ldots,\ell_t$ (as has been done, to some extend, for instance in [4], see also [8]), I prefer the above less conventional, but also less confusing notation.

The group Σ_P and thus the group ring $\mathbb{R}\Sigma_P$ acts from the left on $\mathbb{R}[x_P|P \in P]$ by

$$(\sum_{\sigma\in\Sigma_P} a_\sigma\sigma)(f(x_P|P \in P)) = \sum_{\sigma\in\Sigma_P} a_\sigma f(x_{\sigma(P)}|P \in P).$$

We define f to be a chirality function, if $\chi_S f = f$. As above we define the "phase portrait" $(\mu_\gamma(f)|\gamma$ active for S) to be the family of numbers of irreducible components of Ch(S)f isomorphic to $N_{S,\gamma}$, thus

$$\mu_\gamma(f) = \frac{\mathrm{Dim}_{\mathbb{R}}\ \ell_\gamma\cdot \mathrm{Ch}(S)f}{n_{S,\gamma}} .$$

Now associate to any ligand $\ell \in L$ a real number $\lambda(\ell)$, the "ligand parameter". Set-theoretically λ is just a map $\lambda: L \to \mathbb{R}$. Then a chirality function f and the ligand parameter λ together define a chirality observation $F_{f,\lambda}: M \to \mathbb{R}: m \to f(\lambda(m(P))|P \in P)$, i.e. $F_{f,\lambda}(m)$ is the value of $f(x_P|P \in P)$ at $x_P = \lambda(m(P))$, $P \in P$. From $\chi_S f = f$ one deduces easily $\chi_S F_{f,\lambda} = F_{f,\lambda}$; more generally one has
$(\sigma F_{f,\lambda})(m) = F_{f,\lambda}(m\sigma) = f(\lambda(m\sigma)(P))|P \in P) = f(\lambda(m(\sigma P))|P \in P) = F_{\sigma f,\lambda}(m)$.

In case a given chirality observation F can be described in the from $F = F_{f,\lambda}$ for some chirality function f and some ligand parameter λ, we will say that (f,λ) is a model for F. The most important question in the Ruch/Schönhofer chirality theory is how to construct approriate models for given F. In some way, this is rather an art than a theory, at least, some features like the aesthetics of an "Ansatz" come definitely into play (and aesthetics can relate to truth as much as mathematics), but still, some reasonable remarks can be made.

At first we observe that

$$\mu_\gamma(F_{f,\lambda}) \leq \mu_\gamma(f)$$

for all active partitions γ, since the map $\mathrm{Ch}(S)f \to \mathrm{Ch}(S)F_{f,\lambda}: h \to F_{h,\lambda}$ is a surjective homomorphism of Ch(S) representation spaces. On the one hand, this leads to the notion of "qualitative completeness": f is defined to be qualitatively complete, if $\mu_\gamma(f) = n_{S,\gamma}$ for all active γ. This is obviously a necessary, though not sufficient condition on f to model all sorts of chirality observations. On the other hand, by first splitting F into irreducible components of various types, one may then ask, individually for each component of type γ, for a model (f,λ) with $\mu_{\gamma'}(f) = \delta_{\gamma,\gamma'}$. So, rather than looking for just one qualitatively complete f, one should look for a family $f_{\gamma,i}$ (γ active; $i=1,\ldots,n_{S,\gamma}$) of independent chirality functions with $Ch(S)\cdot f_{\gamma,i} \cong N_{S,\gamma}$. Such families have indeed been constructed by Ruch and Schönhofer in [11], § 12 for a whole lot of instructive examples.

But, to look for the right phase portrait of our model is by far not enough. Of course, if a model (f,λ) for F is found, one wants to interpret this result by interpreting f as the "form" of the observed chiral forces and $\lambda(\ell)$ as the specific parameter of the ligand ℓ, which determines the strength of its interaction. But unfortunately, standard procedures of interpolation of polynomials can be used, to construct for any chirality observation F and any ligand parameter λ with $\lambda(\ell) \neq \lambda(\ell')$ for $\ell \neq \ell'$ a chirality function f with $F_{f,\lambda} = F$. Thus the above interpretation can hold only, if f is not arbitrary, but of a particular "simple" form. Two ways, to define "simplicity" in this context, have been studied by Ruch and Schönhofer. One is, to look for some f of smallest possible degree, the other one is to look for some function $f = f(x_P|P \in P)$ - even if it is not a polynomial in $(x_P|P \in P)$ - which can be written as a sum of functions $f = f_1 + \ldots + f_m$ with $f_i = f_i(x_P|P \in P)$ actually depending on as few as possible of the $(x_P|P \in P)$. In [11] these definitions are treated as "the first and the second approximation method". Both have interesting aspects and consequences. In particular, they are related to certain linear and algebraic identities between the values of the chirality functions, a phenomenon, which has been studied already in various papers (see [5] and [1], for instance) and will be subject to further investigations.

6. Miscellaneous Remarks.

The whole theory is, of course, not restricted to rigid molecules. Once a skeleton S with sites P and a set L of appropriate ligands has been selected, one may consider instead of the rigid symmetry group G_S and its homomorphism $\det: G_S \to \{\pm 1\}$ the permutation-inversion group G of "feasible" permutations and permutation-inversions of the set P of sites (Longuet-Higgins, [7], see also [3]) and its homomorphism $G \to \{\pm 1\}$, whose kernel consists of precisely the permutations in G. Then the rest of the theory works as above,as long as there is some reason to believe that the chirality forces acting in some molecule m are indeed cancelled out up to just one, if m is replaced by the mixture $\psi \cdot m$ with ψ a primitive idempotent in Ch(S).

A good example for this more or less trivial extension of the theory are probably the derivatives of $B(CH_3)_3$.

Another perhaps interesting application of the Ruch/Schönhofer theory might be towards phase transitions in molecules. If F is a chirality observation on the class $M = L^P$, it may depend on various other parameters like temperature, wave length, magnetic fields etc. It may be interesting to study possible changes of the phase portrait of F depending on variations of these parameters.

References

[1] G. Derflinger und H.Keller: Zur Theorie der Chiralitätsfunktionen, V: Zum Konzept der qualitativen Vollständigkeit - eine Kritik, Wien 1977, preprint.

[2] A.Dress: Eine Bemerkung zur Ruch-Schönhofer'schen Halbordnung von Young-Diagrammen. To be published in match, 1979.

[3] A.Dress: Some suggestions concerning a geometric definition of the symmetry group of non-rigid molecules. 1979, these proceedings.

[4] J.Dugundji, D.Marquarding and I.Ugi: The Indeterminateness of Chirality Functions in Hyperchiral Families. Chemica Scripta, 1977, 11,17-24.

[5] W.Hässelbarth, B.Richter und E.Ruch: Über die Güte semi-empirischer Ansätze für Chiralitätsfunktionen, Preprint, Berlin, 1978.

[6] B.Lesche und E.Ruch: Information extent and information distance, J.Chem.Phys.69, 393-401, 1978.

[7] H.C.Longuet-Higgins: The symmetry groups of non-rigid molecules, Molec.Phys.,6, 445-460, 1963.

[8] C.A.Mead: Comment on "Hyperchirality", Chemica Scripta, 1967,10, 101-104.

[9] A.Mead und E.Ruch: The Principle of Increasing Mixing Character and Some of Its Consequences, Theoret.Chim. Acta (Berl.), 41,95-117, 1976.

[10] E.Ruch: The Diagram Lattice as a Structural Principal, Theoret. chim. Acta (Berl.) 38, 167-183, 1975.

[11] E.Ruch und A.Schönhofer: Theorie der Chiralitätsfunktionen, Theoret. chim. Acta (Berl.),19,225-287, 1970.

[12] D.Haase und E.Ruch: Quanten-mechanische Theorie der optischen Aktivität der Methanderivate im Transparenzgebiet, Theoret.chim. Acta (Berl.), 29, 189-234, 1973.

[13] D.Haase und E.Ruch: Quanten-mechanische Theorie der optischen Aktivität der Allenderivate im Transparenzgebiet, Theoret.chim. Acta (Berl.),29, 247-258, 1973.

[14] G.Kresze, E.Ruch und W.Runge: Experimentelle Prüfung von Näherungsansätzen für Chiralitätsfunktionen am Beispiel der optischen Aktivität von Allen-Derivaten im Transparenzgebiet, Angew. Chemie, 85, 10-15, 1973.

[15] B.Richter, W.J.Richter und E.Ruch: Experimentelle Prüfung von Näherungsansätzen für Chiralitätsfunktionen am Beispiel der optischen Aktivität von Methan-Derivaten im Transparenzgebiet, Angew. Chemie, 85, 21-27, 1973.